Seaports and Development in
Tropical Africa

Seaports and Development in Tropical Africa

EDITED BY

B. S. Hoyle, M.A., Ph.D., F.R.G.S.

Lecturer in Geography,
University of Southampton
(formerly Makerere University College, Uganda
and University College of Wales, Aberystwyth)

AND

D. Hilling, M.Sc.

Lecturer in Geography,
Bedford College, University of London
(formerly University of Ghana)

Palgrave Macmillan

First published 1970 by
MACMILLAN AND CO LTD
Little Essex Street London WC2
and also at Bombay Calcutta and Madras
Macmillan South Africa (Publishers) Pty Ltd Johannesburg
The Macmillan Company of Australia Pty Ltd Melbourne
The Macmillan Company of Canada Ltd Toronto
Gill and Macmillan Ltd Dublin
ISBN 978-0-333-11217-5   ISBN 978-1-349-15362-6
DOI 10.1007/978-1-349-15362-6

Contents

List of Plates   vii

List of Figures   ix

List of Tables   xi

Preface   xiii

Foreword by Sir Arthur Kirby, K.B.E., C.M.G.   xv
(Chairman, National Ports Council)

1   **Seaports and the Economic Development of Tropical Africa**   1
D. *Hilling* (Bedford College, University of London)
B. S. *Hoyle* (University of Southampton)

2   **The Morphological Development of West African Seaports**   11
H. P. *White* (University of Salford)

3   **Nouadhibou (Port Étienne) and the Economic Development of**   27
Mauritania
Charles Toupet (University of Dakar)

4   **The Changing Role of the Port of Dakar**   41
Assane Seck (Ministry of Education and Culture, Government
of Senegal, and University of Dakar)

5   **Physical Potential and Economic Reality: the Underdevelop-**   57
ment of the Port of Freetown
J. *McKay* (University College of Dar es Salaam)

6   **The Ports of Liberia: Economic Significance and Development**   75
Problems
W. *Schulze* (University of Giessen)

7   **The Development of the Port of Abidjan and the Economic**   103
Growth of Ivory Coast
Mireille Bouthier (University of Abidjan)

8   Port Development and Economic Growth: the Case of Ghana   127
    D. Hilling (Bedford College, University of London)

9   Cotonou: Some Problems of Port Development in Dahomey   147
    A. Mondjannagni (University of Abidjan)

10 Patterns and Problems of Seaport Evolution in Nigeria   167
    Babafemi Ogundana (University of Ife)

11 Problems of Port Development in Gabon and Congo-   183
    Brazzaville
    Pierre Vennetier (University of Bordeaux)

12 The Development and Problems of Port Sudan   203
    M. M. Khogali (University of Khartoum)

13 The Emergence of Major Seaports in a Developing Economy: 225
    the Case of East Africa
    B. S. Hoyle (University of Southampton)

14 Ports and Economic Integration in Madagascar   247
    William A. Hance (Columbia University)

    Index   267

List of Plates

1 **Nouadhibou (Port Étienne):** The iron-ore loading pier
(*C. Toupet*)

2 **Dakar:** The southern zone of the port
(*Photo Artis, Dakar*)

3 **Monrovia:** General view of the port
(*Professor Schulze*)

4 **Buchanan:** General view of the port
(*Professor Schulze*)

5 **Abidjan:** Timber loading in Banco Bay
(*Photo Ciné, Abidjan*)

6 **Abidjan:** View across the fishing harbour, West Quay and
Plateau area of town
(*Photo Ciné, Abidjan*)

7 **Accra:** Surf-boat loading prior to 1962
(*J. Allan Cash*)

8 **Takoradi:** General view of port
(*J. Allan Cash*)

9 **Tema:** The Volta Aluminium Company's berth and alumina
store
(*Volta Aluminium Company*)

10 **Cotonou:** The lagoon entrance, old pier and new port
(*Port of Cotonou*)

11 **Port Harcourt:** General view of port
(*Shell Company*)

12 **Lagos–Apapa:** The new container berth
(*D. Hilling*)

13 **Pointe Noire:** General view of port and new extensions
(*P. Vennetier*)

14 **Mombasa:** Port and industrial area
(*East African Railways and Harbours*)

15 **Dar es Salaam:** The town and main quays
(*East African Railways and Harbours*)

List of Figures

2.1   *West Africa: the nature of the coastline, and the pattern of port development*   13

3.1   *Economic map of Mauritania*   28

3.2   *The site of Nouadhibou (Port Étienne)*   29

4.1   *Layout and facilities of the port of Dakar*   42

4.2   *The hinterland of the port of Dakar*   47

4.3   *The site of Dakar*   53

5.1   *Sierra Leone: main centres of economic activity*   63

5.2   *The ports of Freetown and Pepel*   65

6.1   *Economic map of Liberia*   76

6.2   *The port of Monrovia: layout and facilities*   81

6.3   *The port of Buchanan: layout and facilities*   82

6.4   *The port of Greenville*   83

6.5   *Cargo traffic at the free port of Monrovia, 1950–67*   86

6.6   *The port of Buchanan: monthly variations of rainfall, iron-ore shipments, and movements of ore-carriers, 1964–7*   98

7.1   *Location of the port of Abidjan*   104

7.2   *The port of Abidjan: layout and facilities*   107

7.3   *Ivory Coast: patterns of export crop production and surface transport*   113

7.4   *Growth of port traffic at Abidjan, 1951–67*   120

8.1   *Economic map of southern Ghana*   132

8.2   *The port of Takoradi*   135

8.3   *The port of Tema*   141

9.1   *Location of the port of Cotonou*   148

9.2   *The port of Cotonou: layout and facilities*   155

10.1   *Functional relationships in the Nigerian port complex*   168

11.1   *Equatorial Africa: surface transport facilities and major mineral resources*   184

11.2   *Seaports of Gabon and Congo-Brazzaville*   188

11.3   *The ports of Port Gentil and Pointe Noire*   194

11.4   *Growth of port traffic at Pointe Noire*   196

12.1   *Port Sudan: layout and facilities*   207

12.2   *The railway system of the Sudan*   213

13.1   *Some environmental factors influencing the emergence of East African seaports*   229

13.2   *Some examples of East African seaport hierarchies*   233

13.3   *Some economic factors involved in the growth of East African seaports*   241

14.1   *International and coastwise traffic at the ports of Madagascar in 1965*   253

14.2   *The road system of Madagascar*   255

14.3   *The commercially productive regions of Madagascar*   258

NOTE: Unless otherwise indicated, maps are oriented north–south.

List of Tables

3.1   *Air traffic at Nouadhibou, 1960–7*   32

3.2   *Cargo traffic handled at Nouadhibou, 1961–8*   36

3.3   *Exports of iron ore from Point Central (Nouadhibou), 1963–8*   38

5.1   *Freetown: exports of major agricultural crops, 1900–45*   60

5.2   *Growth of Freetown's port traffic, 1824–1953*   61

5.3   *Growth of Freetown's port traffic, 1961–5*   62

5.4   *Composition of Sierra Leone's export trade, 1965–7*   64

5.5   *Sierra Leone: rough and uncut diamonds purchased by Government Diamond Office, 1964–7*   67

5.6   *Sierra Leone's balance of payments position, 1964–7*   68

5.7   *Structure of Sierra Leone's import trade, 1965–7*   69

6.1   *Geographical characteristics of Liberian ports and their hinterlands*   78

6.2   *Technical characteristics of Liberian ports*   80

6.3   *Ranking of Liberian ports by percentage share of total traffic*   84

6.4   *Ranking of Liberian ports in order of exports*   85

6.5   *Cargo movements at the free port of Monrovia, 1950–67*   87

6.6   *Number of ships calling at the free port of Monrovia, 1958–67*   87

6.7   *Iron ore shipped and number of ore-carriers calling at the port of Buchanan, 1964–7*   89

6.8   *Destination of iron ore shipped from the port of Buchanan, 1963–6*   90

6.9   *Port development in Liberia: a geographical and technical analysis*   100

7.1   *Value of principal commodities exported through the port of Abidjan, 1951–66, by percentages of total exports*   122

7.2   *Value of imports at Abidjan in 1953 and 1966: percentage distribution between major sectors*   122

7.3   *Traffic of the port of Abidjan in 1953 and 1966, by commodities and by countries and monetary zones*   124

7.4   *Value of the external trade of Ivory Coast shown as a percentage of the total trade of former French West Africa, 1947–58*   125

8.1   *Ghana's port traffic, 1938 and 1945–67*   130

8.2 *Takoradi port traffic, 1937–8 to 1963–4*   137

8.3 *Trade structure of Tema and Takoradi, 1964–7*   144

9.1 *Traffic at the port of Cotonou, 1948–67*   157

9.2 *Passenger traffic at Cotonou, 1959–67*   160

9.3 *Traffic of the port of Cotonou with France and with the franc*   160
zone as a percentage of the total traffic of the port, 1959–67

9.4 *Port of Cotonou: projections for Dahomey and Niger*   162
traffic, 1965–85

10.1 *The pattern of seaport evolution in Nigeria*   172

12.1 *External trade of the Sudan: percentages of the total volume*   204
handled by Port Sudan, Wadi Halfa and the frontier posts
(including Khartoum airport), 1938–67

12.2 *Number and tonnage of vessels entering Port Sudan, 1908–24*   210

12.3 *Number and net registered tonnage of vessels entering Port*   211
Sudan, 1925–50

12.4 *Shipping and cargo traffic handled at Port Sudan, 1951–2*   212
to 1967–8

12.5 *Port Sudan: volume of principal commodities handled, 1962–3*   215
to 1967–8

12.6 *Direction of Sudanese external trade by percentage of value,*   216
1954 and 1967

12.7 *Some indices of congestion at Port Sudan, 1954–67*   218

12.8 *Port Sudan: tonnage of goods imported and railed up*   220
country during the 1966–7 and 1967–8 seasons

13.1 *Facilities and equipment at East African seaports*   226

13.2 *East African seaports: shipping, passengers and cargo dealt*   227
with in 1968

13.3 *Percentage of total cargo traffic at East African mainland sea-*   243
ports handled at individual terminals, 1956–68

13.4 *Some examples of factors involved in the emergence of the*   244
modern East African seaport group

14.1 *Madagascar exports by value and percentage of total value,*   249
for selected years 1938–67

Preface

The essays in this volume have been brought together to emphasise the important role of seaports in the economic growth of underdeveloped areas, and to demonstrate the increasing interest now focused upon this problem by geographers in tropical Africa. The essays are introduced in the first chapter; whilst each contributor draws attention to the problems of a specific port or port group, it should be stated that many of the contributors are familiar at first hand with the problems of a wide range of ports in Africa and elsewhere, and that during the past five years the editors have between them visited almost all the ports discussed in this volume. Nevertheless we have not sought to provide a systematic coverage of tropical African seaports, but rather to bring together a series of illustrations and analyses of representative port problems. Lack of reasonably up-to-date information (for example on ports of Portuguese Africa) has also encouraged a selective approach.

It is a pleasure to place on record our gratitude to the contributors for their full co-operation, and to the many port and other authorities in tropical Africa and elsewhere without whose aid the research on which these essays are based could not have been completed. We acknowledge also the willing assistance of Mrs Shirley Loates, B.A., and Mrs Norma Tatum, B.A., who helped with the translation of four of the five chapters originally submitted in French. Most of the maps and diagrams were redrawn and reproduced by Mr M. Hughes, Mr D. Griffiths, Mr M. Gelly-Jones and Miss T. J. Kinsey (members of the cartographic and technical staff of the Department of Geography, University College of Wales, Aberystwyth); Mrs Mair Jenkins and Miss Carol Parry (of the same department) typed the final version of several chapters. Miss Juliet Williams, B.A., former Geography Editor of Messrs Macmillan & Co. Ltd, was most helpful during the initial planning stages; her successor, Miss Anne Ralphs, and her colleagues, have dealt most efficiently with the later stages of production.

Finally, we are indebted to Professor James Bird, Professor William A. Hance, Professor R. J. Harrison Church, Professor Guy Lasserre, Professor J. Oliver and Professor R. W. Steel for their helpful advice, comment and criticism.

B. S. HOYLE
University of Southampton
D. HILLING
Bedford College, University of London

March 1969

Foreword

Sir Arthur Kirby, K.B.E., C.M.G.

Chairman, National Ports Council

This volume of essays provides a consensus of information which will be of great value to geographers and economists interested in tropical African transport. To me, at a time when we are seeking to determine a pattern for port development in Great Britain, these essays demonstrate vividly that ports are developed to satisfy a variety of requirements – administrative demands, political ambitions, or the trading needs of the time – without regard to any overall national or regional design. Tropical Africa is not peculiar in this respect; the same can be said about shortcomings in efficiency in ports not covered by these essays, for most of the inefficiencies stem from causes which are world-wide. Hence the distortions which can be seen against a pattern of desirable traffic flows are no more serious in tropical Africa than elsewhere – as for example in Britain, where the port pattern survives from the heyday of massive coal exports and includes a legacy of several ports that by today's standards should not exist, but which for social, political and economic reasons are difficult to close.

When I first knew West Africa, forty years ago, the only means of transferring cargo between sea and land along the exposed coast of the Gulf of Guinea was by small boats carrying only about two tons and manned by six or eight men paddling hard to span the distance of a mile or more between the ship at anchor and the surf-beaten shore. Passengers were put ashore in an open box-like affair holding four people, called a 'mammy' chair, which was slung overside by a ship's derrick into a small surf boat heaving up and down in a 10 to 20 ft swell. Only the skill of the winch-men could ensure that the 'mammy' chair hit the surf boat at the right moment, and I have known cases where unpopular officials were 'accidentally' tipped into the sea! Today, although lighterage ports are still commonplace, the major seaports of tropical Africa are all well equipped with deep-water facilities. Because these facilities have been developed during the present century they are fairly modern by world standards, and tropical African ports are often less encumbered than ports elsewhere with outmoded facilities difficult and expensive to adapt to modern requirements.

Remedies for inefficiencies in port operation are seldom within the control and competence of port managements. It is not sufficiently recognised that the efficiency of ports is critically dependent upon the users – shippers, receivers and shipowners alike. Ports cannot be treated in isolation. Many users appear to regard them as rectifiers in the transport chain, and expect ports to smooth out the consequences of inefficient trading practices and to restore tidiness to the unco-ordinated and badly managed movement of cargo by land and sea. They also expect ports to absorb peaks of agricultural production and fluctuations of movement arising from market-price variations. An examination of congestion arising from cocoa or groundnut movements in West Africa or from coffee and cotton exports in East Africa will illustrate this.

In spite of these problems many of the seaports of tropical Africa are efficient by world standards and adequate provision exists for the movement of bulk cargoes such as oil and ores; it is in the field of general cargo movements that difficulties arise. No matter how modern the port facilities may be, congestion and increased costs will be incurred unless general cargo movements are fully co-ordinated. An example of existing co-ordination is provided by bulk oil, which is the most efficiently moved cargo because it is controlled by one organisation throughout transit from the field to the refinery, and is transferred between ship and shore in one unbroken operation. The movement of general cargo will become more efficient only when it is bulked into containers, or other unitised forms, moving smoothly from an inland port to the ship, and vice versa, without break of bulk at the ports. It is the users, the shipper and the shipowner, who have to take the decisions to bring this about. The techniques for moving general cargo in bulk which are now being applied on the North Atlantic routes, and in Europe and the Far East, will soon be seen also in tropical African ports.

A recent report to the United Nations Economic Commission for Africa suggests that an Advisory Office for the West African sub-region should be set up to organise on a regional basis such matters as port management, cargo handling, mechanisation, documentation, labour training, tariffs, shipping legislation, etc. Such a system would be of great value to all the countries concerned, and would go far towards co-ordinating cargo movements by sea and land. It would represent an important step towards a greater measure of economic integration amongst the countries of modern Africa, which is widely regarded as one of the continent's most pressing needs.

1. Seaports and the Economic Development of Tropical Africa

D. Hilling and B. S. Hoyle

'Intensive utilisation of the tremendous potential resources of that continent (Africa) is certain to develop during the post-war years, and trade will expand manyfold. Consequently, the harbours of Africa will lose their former provincial status and will become a matter of international concern.'[1]

In retrospect the 1960s may be identified as a period of major innovation in maritime transport. The ever-increasing size of bulk carriers and the unitisation of general cargo are trends which are clearly gaining in momentum and which will leave few ports unaffected. The time is therefore opportune for economic geographers to analyse the existing port systems and to assess the potential for change. Nowhere is this more important than in the developing countries of the tropical world where the ports assume such a critical role in the development process.

It has been claimed for transport that is 'the formative power of economic growth and the differentiating process',[2] and that 'a lack of transport facilities is . . . a major deterrent to rapid economic growth and social progress. Transport difficulties have considerably retarded the exploitation of natural resources, industrialisation, expansion of trade . . . and in some cases the achievement of national unity.'[3] What is true of transport in general is certainly true, in this context, of seaports in particular. At the present time 94 per cent[4] of the total external trade of the African continent passes through its ocean ports; trade is orientated overwhelmingly towards overseas countries, partly as a result of the long period of colonial dependence which most parts of Africa have experienced, but also because of a marked similarity of resource endowment which

1 G. F. Deasy, 'The Harbours of Africa', Economic Geography, *vol. 18 (1942)* 325–42.
2 F. Voigt, The importance of the Transport System for Economic Development processes, *United Nations Economic Commission for Africa*, E/CN. 14/CAP/3G (*Addis Ababa, 1967*).
3 *United Nations Economic and Social Council*, Transport Development, E/4304 (*New York, 1967*).
4 *United Nations Economic Commission for Africa*, A survey of Economic Conditions in Africa, E/CN. 14/397 (*Addis Ababa, 1967*).

makes many of the countries of modern Africa competitors for overseas markets rather than natural trading partners. Development of overseas trade is a decisive factor stimulating the emergence of the money economy, the expansion of urban populations and the growth of local markets for goods and services; trade also helps to create conditions favourable to the development of modern agricultural systems and the establishment of industry.[1] Throughout Africa, the provision of port facilities has been a necessary precondition of modern economic growth; today, the stage of economic development reached in a given part of the continent is in considerable measure a function of the capacity and degree of sophistication of the port facilities available.

In bringing together these essays the editors acknowledge that the ports of the developing countries of tropical Africa share certain common characteristics which distinguish them from ports serving countries with more advanced economic systems. Of paramount importance has been the inheritance from the colonial era: with few exceptions, the countries of tropical Africa have spent varying but considerable periods under colonial control, and even a country such as Liberia (an independent republic in 1847) shares the essential economic characteristics of the colonial and former colonial territories. The trade structure which normally developed in African territories under European control was highly imbalanced, characterised chiefly by the exportation of a limited range of agricultural and mineral primary products and the importation of a varied assortment of semi-processed and manufactured goods. Not only was the trade in either direction of a fundamentally different nature, but the volume of exports almost invariably far exceeded that of imports. This dual imbalance is the primary characteristic of what is sometimes termed a 'colonial' trade structure; it was, and is, a natural reflection of capital investment in enterprises involving least risk, for neither the size nor the purchasing power of local African markets could support the establishment of manufacturing industry on a considerable scale.

The colonial legacy in terms of the transport infrastructure itself is also of major importance. African colonial territories certainly started out along the road leading to full integration in the modern world economy whilst their affairs were still in European hands, but the overall approach of the colonial powers towards economic development tended to be passive and lacking in positive content; it was only when it became clear that the days of European rule in Africa were numbered that colonial

---

1 S. D. Newmark, Foreign Trade and Economic Development in Africa (Stanford, 1967).

authorities adopted a more aggressive approach towards the economic and social advancement of their dependencies. Transport infrastructures were initially developed frequently for military, political or administrative reasons rather than for economic motives, and even where the economic motive was uppermost transport was designed with the export of a limited range of primary products in mind. A system based on such considerations is not suited either to the internal exchange of locally produced goods or to the distribution of imports from overseas. 'Almost nowhere in tropical Africa is there a fully integrated transport complex permitting a rational selection of either road, rail or air transport.'[1] The arbitrary nature of the colonial political division of Africa, which has now become the division into independent states with few changes of boundary location, meant that transport systems were developed largely in isolation to serve the needs of particular political units rather than the natural trading areas or hinterlands. Few of the territories of colonial Africa had a port system capable of supporting rational, co-ordinated economic growth within the framework of territorial divisions. The gradual emergence in tropical Africa of a number of major seaports at the coastal termini of railways penetrating in many cases far into the interior, but with no interconnection between the railways, meant not only that vast areas of the continent were inadequately served but also that economic growth tended to be highly concentrated in favoured coastal and interior 'islands'.[2]

The immediate post-war period was characterised by an almost insatiable demand for the raw materials of tropical Africa, and in Africa itself there was the need to make good the deprivations of the war years. The total volume of external trade of Britain's West African territories increased by 90 per cent in the ten years following the war, and between 1945 and 1949 the external trade of French West Africa trebled. With the approach of political independence there was a new urgency to stimulate economic growth.

The financing of economic development must to a considerable extent depend upon revenue derived from the export base, and during the past decade there have been considerable increases in output in the main agricultural and mineral sectors. To some extent the traditional imbalance in the trade structure of African countries has been further emphasised. In 1960 bulk cargoes accounted for 7·4 and 41·0 per cent respectively of the value and volume of total African exports, but only four years later the

1 W. A. Hance, African Economic Development, 2nd ed. (New York, 1967).
2 W. A. Hance, V. Kotschar and R. J. Peterec, 'Source Areas of Export Production in Tropical Africa', Geographical Review, vol. 51 (1961) 487–99.

figures had increased to 19·0 and 72·0 per cent.[1] Many of these bulk products are characterised by widely fluctuating world prices and inelasticity of demand, and their considerable bulk and low value implies that transport costs make up a high percentage of the delivered price. It follows therefore that if a tropical country is to retain a competitive position in bulk produce, the cost of transport must be reduced to a minimum. Recent United Nations studies[2] have demonstrated the importance of port charges in the overall determination of freight rates; port charges themselves are influenced by a variety of factors, but the efficiency and capacity of the facilities provided in the ports and the ability to effect rapid ship turnround is of critical importance.

Questions of port efficiency and charges are also fundamental in the context of the highly varied and rapidly increasing import traffic which most tropical African ports handle. Whilst a 1 per cent increase in the per capita income of a developed country increases the demand for raw materials and food by only 0·6 per cent, the same per capita increase in a developing country will result in a 1·8 per cent increase in imports.[3] A greater volume of exports results in increased local purchasing power for both individuals and governments, and leads to a rapid expansion of the volume of imports and an increase in demand for both capital and consumer goods. In many countries the gap between export and import tonnages has been narrowed, and port facilities, geared as they have traditionally been to exports rather than imports, have been placed under severe strain, many having been shown to be quite unable to support high-level economic growth.

The achievement of political independence throughout the greater part of tropical Africa has stimulated new efforts to attain economic independence and has created an urgent need to establish the preconditions essential for economic 'take-off'. Demands for economic diversification, for a reduction in dependence upon a narrow range of primary products, and above all for industrialisation represent a new climate of thought on African economic problems and have stimulated considerable improvements in the transport infrastructure of the region as a whole. The extension and elaboration of individual ports and regional port systems

1 *United Nations Economic Commission for Africa, op. cit.*
2 *United Nations Economic Commission for Africa*, Preliminary Survey of Factors Contributing to the Level of Freight Rates in the Seaborne Trade of Africa and Related Matters, *Part I:* The West and Central African Sub-regions, E/CN. 14/TRANS/27; *Part II:* The East African Sub-region, E/CN. 14/TRANS/27 (*Addis Ababa, 1966*).
3 *A. F. Ewing*, Industry in Africa (*London, 1968*) p. 1.

has been a major element in this process. This general economic and political situation distinguishes port developments in the tropics from those in more advanced countries where port growth generally takes the form of gradual evolution, with the result that the ports inherit centuries of experience and the advantage of hinterlands with a high technological level and relatively abundant supplies of capital for investment. The less developed countries have been faced with a sudden need to modify existing ports or to establish new facilities with none of the advantages of a long-established and mature economy, with the possible exception that they are not too encumbered with bureaucracy, vested interests and congested hinterland conditions.[1] It follows that the less developed countries cannot afford to make mistakes; for each port development project will have a far greater impact on the economic geography of the hinterland it serves than an equivalent development at a port in a mature economy, and the nature of the port facilities provided will directly influence the pattern and direction of economic advance in the country concerned. Thus for many years to come the pattern of economic development in tropical African countries will be in part a function of the port facilities now being created or expanded, and in this way port development becomes a vital sector in national physical and economic planning. Developing countries must be prepared to break the vicious circle in which port developments are postponed until increased traffic provides an obvious economic justification but traffic cannot be further increased until port facilities are extended.

The life expectancy of major port installations is considerable, certainly of the order of fifty to a hundred years, but the present rapid changes in the technology of maritime transport make decision-making difficult and not without risk whatever the existing level of national economic attainment. For a developing country to ignore this new technology would certainly mean a widening of the gap which separates the developed and less developed nations, not only in the sphere of transport but over the whole field of economic development. However, the fact that at least some of the developing countries are establishing their basic port infrastructure during a period when maritime transport is itself undergoing vitally important changes undoubtedly adds to the problems facing port authorities but also provides splendid opportunities for the incorporation of new methods at an early stage in port growth.

This is the broad context within which these essays have been written.

---

1 B. Nagorski, '*Port Problems in Developing Countries*', Dock and Harbour Authority, *vol. 49, no. 572 (1968) 36–43.*

The object of this volume is not to provide a systematic account of the various individual ports and port groups of tropical Africa, but rather to focus attention, against the geographical and economic background, upon some of the diverse problems encountered in the attempt to create viable, rational port systems within the region. The geographer, with his interest in the 'integrations composed of interrelated phenomena of the highest degree of heterogeneity',[1] cannot but be fascinated by the seaport and by the complex interplay of physical, geographical and socio-economic phenomena it represents; but from the very complexity of seaports it follows that 'differences in the theory and method of the study of ports are reflected by the number of approaches to the subject that have been used by geographers'.[2] The variety of approach represented by this volume stems in part from the diversity of the subject matter, although other studies have placed greater emphasis upon similarities in the evolving transport patterns in underdeveloped countries.[3] The basic problem of all ports is similar, namely the rapid, safe and economical transhipment of cargo,[4] but these essays show clearly that the solution of this problem in the tropical African context, in spite of the common characteristics described above, varies greatly from port to port. Generalisations are nowhere more dangerous than in the tropical African area, where rapidly changing conditions and the variety and complexity of the operative factors implies that every port is a separate case for individual study. To some extent, therefore, this volume is based upon studies of individual terminals, but broader aspects are not neglected and some attempt is made to indicate the nature and range of problems encountered and to demonstrate a variety of individual solutions.

In common with geographers working in other spheres, the port geographer is concerned with the influence of the environment upon economic activity. A port reflects in its layout the influence of its immediate land and water site, whilst in functional terms a stronger force is exerted by its wider relationships with the hinterland and foreland areas.[5] The varied environments of tropical Africa offer a range of physical problems, some of them peculiar to the region, which affect port development in

1 R. *Hartshorne*, Perspective on the Nature of Geography (*Chicago, 1959*) *p. 35.*
2 *J. N. H. Britton*, '*The External Relations of Seaports*', Tijdschrift voor Economische en Sociale Geografie, vol. *56 (1965) 109–12.*
3 *E. J. Taafe, R. L. Morrill and P. R. Gould*, '*Transport Development in Under-developed Countries: A Comparative Analysis*', Geographical Review, *vol. 53 (1963) 503–29.*
4 *Nagorski, op. cit.*
5 *J. Bird*, The Major Seaports of the United Kingdom (*London, 1963*).

important ways. Broadly, Africa's coasts are not well endowed with natural harbours[1] of adequate depth and ease of access for modern navigation, and the problems of creating artificial ports are great although not insuperable. Reference is made by many contributors to this volume to specific problems of coastal morphology and hydrography affecting port growth; particular attention is given to the problem of the surf barrier and littoral sand movement in West Africa (Bouthier, Mondjannagni) and to the coral hazard along the ria coastline of eastern Africa (Hoyle, Khogali); the relationship between the evolving morphology of the seaport and that of the coast is illustrated in detail (White); the wide range of physical problems encountered is illustrated by the essays which discuss Nouadhibou (Toupet) and the Liberian ports (Schulze); Mondjannagni and Bouthier both discuss the problems of sand movement and accumulation along the lagoon coast of the Gulf of Guinea, and show how the ports of Cotonou and Abidjan have adopted contrasting solutions delicately adapted to local environmental conditions.

The use of the term 'hinterlands' by port geographers is commonplace and capable of much refinement, but nevertheless essential to an understanding of the functional aspects of any port. The relationship between a port and 'the organised and developed land space'[2] which it serves is highly complex and very intimate but often defies precise delimitation. The utilisation of the resources of the land area influences the demand for port facilities, and the provision of facilities is in turn a vital factor in the development and prosperity of the hinterland. In the developing countries the relationship between port and hinterland is perhaps somewhat clearer than in more advanced areas of the world; since links between the various colonially induced port–railway systems are few, ports tend to assume undisputed control over the external commodity flow of a particular area. Thus a 'primitive hinterland'[3] develops, sometimes involving the outflow of a single major commodity such as iron ore, and in such cases the port–hinterland system is of little general significance.

Another factor of undoubted importance in the context of tropical African port hinterlands is the political situation; the evolution during the colonial era of transport infrastructures and port systems to serve arbitrarily defined territories is paralleled today by the creation of new

1 *Deasy, op. cit.*
2 G. Weigend, '*The Problem of Hinterland and Foreland as Illustrated by the Port of Hamburg*', Economic Geography, *vol. 32 (1956) 1–16.*
3 F. W. *Morgan*, Ports and Harbours (*London, 1958*).

ports designed to serve limited national hinterlands derived from the colonial pattern. The almost simultaneous development of new seaports in Dahomey and Togo is a case in point, whilst the West African colonial port system was largely artificial, and only in the context of political independence and increasing inter-state economic co-operation is a more natural system based upon trade corridors beginning to emerge. In East Africa, as Hoyle shows, the emergence of the modern port group which now serves the area has been influenced by a wide range of factors inter-acting with one another, but the changing economic and political geography of the region as a whole has been of paramount importance. A special problem associated with port hinterlands is that of the landlocked state, of which there are numerous examples in tropical Africa today; this inevitably creates serious problems of traffic flow, hedged around by international agreements and a variety of political motivations. Seck discusses the role of Dakar in relation to Malian trade, and Vennetier considers the importance of Pointe Noire to the Central African Republic and Chad.

Many of the countries of tropical Africa have convenient access to only one major seaport. This presents serious problems of congestion in many cases, but problems of a different kind arise where a country is served by several terminals. McKay illustrates how a small and poorly developed national hinterland in Sierra Leone, equipped at Freetown with one of tropical Africa's finest natural harbours, may nevertheless utilise a variety of alternative outlets. Schulze examines the case of the Liberian ports in the context of the actual and potential development of their separate hinterlands, and Hance demonstrates how a proliferation of small, ill-equipped ports in Madagascar serving restricted hinterlands has led to economic sectionalism, problems of economic and political integration and a misplaced emphasis on Tananarive.

The multifunctional character of a seaport extends to the industrial as well as the commercial activities associated with the transfer of goods. Industrial activity within the port zone itself is a common feature in tropical Africa, as is the growth of port-orientated industries at no great distance from the quayside but requiring some intermediary transport; the development of fishing harbours within the confines of modern sea-ports is also frequently important, and some commercial ports operate in addition as naval bases. The functional structure of a port in commercial, industrial and military terms is determined partly by the nature of the hinterlands and partly by the water situation of the port in relation to overseas maritime links; moreover, different port functions relate to different areas of the hinterlands, on an international, national or purely local scale. Because ports are multifunctional they tend in developing countries to become the growth poles at which economic development is concentrated, and a high percentage of new industry in those tropical African countries with a seaboard is located in the ports. The extent to which a seaport acts as a growth pole or, conversely, as a factor restricting economic advancement is discussed in Hilling's essay on Ghanaian seaports.

The relationship between port development and the economic and social advance of the independent countries of tropical Africa is a recurring theme in this volume. The seaport emerges as a vital factor in the economic growth of this part of the less developed world; permissive rather than positively stimulative, the economic implications of the seaport's multi-functional role touch almost every aspect of the life of the populations served. The provision and improvement of port facilities cannot therefore be a matter of chance, but must become a high-priority concern within the sphere of national and regional economic planning.

2. The Morphological Development of West African Seaports

H. P. White

Mr White is Reader in Geography at the University of Salford where he specialises in transport. He worked in West Africa between 1952 and 1961 where he became interested in port development, publishing a number of papers on the subject. In 1967 he was a member of a team advising the Nigerian Government on traffic trends through their ports in relation to investment.

In West Africa port development is of vital concern to the thirteen states and one remaining colonial territory into which it is divided, as much to the three landlocked countries as to those with a coastline. This is for two reasons. Communications across the landward boundaries of West Africa are so poor and difficult that trade is negligible. Secondly, all West African countries are dependent on overseas trade for sustaining their ambitious programmes of economic and social development. Apart from any direct foreign aid, it is by the expansion of exports that they can finance the imports of capital goods and other commodities, such as cement, metal products and petroleum, needed to equip and sustain the expanding infrastructure and manufacturing industries. Thus the period since 1945 has been characterised by a marked increase in the tonnage of bulky raw materials exported and an even more spectacular increase in the volume of imports.

Port capacity has therefore been a limiting factor in economic development. Saturation of facilities at Takoradi harbour was considered to be a root cause of the severe inflation which affected Ghana between 1950 and 1952[1]. At Abidjan there was a constant struggle throughout the 1950s to increase port capacity to keep pace with the increasing volume of overseas trade in Ivory Coast, but even so congestion remained.[2] Finally, it was said that at Cotonou 'le wharf est devenu un véritable goulôt d'étranglement pour les économies du Dahomey et du Niger'.[3] Costs of port

1 D. Seers and C. R. Ross, Report of the Financial and Physical Problems of Development in the Gold Coast (*Accra, 1952*).

2 J. L. Tournier, '*The Port of Abidjan*', Dock and Harbour Authority, *vol. 39 (1958) 371.*

3 M. Sharlet, '*Le Bénin-Niger*', Vie du Rail d'Outre-Mer (*Dec 1960*) pp. 3–8.

operation in general and of congestion in particular have to be considered in the context of export costs for very competitive markets, but capital investment in port facilities looms very large in total investment in the infrastructure.

A study of ports is therefore necessary to an understanding of the West African economy. In this essay, however, there is no intention of analysing the trade of the ports, or the extent of their hinterlands. It is proposed instead to make a brief examination of the form and equipment of the ports and to indicate how this is influenced by various physical and economic factors. Because of these factors, the morphological development of West African ports shows a discernible pattern. This idea, similar to that propounded by Bird in tracing the development of 'anyport'[1] was first put forward by the present author in 1959,[2] but in view of the rapidity of subsequent development it is worth redefinition.

The nature of the coastline

The principal causative factor is the nature of the coastline (Fig. 2.1). Under natural conditions this functions more as a barrier than as a zone of contact between land and sea. This coast, extending for some 3,840 km. (2,400 miles) between Cape Verde and Mount Cameroun, is for the most part fault-guided and generally lacks major indentations. The contrast with the coast of Atlantic Europe or of Japan is extreme. Between Cape Verde and Cape St Anne (Sierra Leone), however, there are large and spectacular rias, but, on the other hand, on the southern coast they are much smaller.

Throughout the year the Atlantic swell breaks on the shore as a very heavy surf. This surf varies in amplitude seasonally, but heights of up to 2·5 m. (8 ft) have been recorded.[3] North of Sierra Leone the direction varies, the south-westerly swell giving way to a north-westerly one between October and May. The junction of the North and South Atlantic swells is sometimes plainly visible, and in June the author has seen surf breaking with equal severity on the north-west and south-west facing beaches of the Cape Verde peninsula. Unless protection can be given from the swell, ships cannot tie up to a quay, but must lie offshore and transfer their cargoes to small boats of special design and limited capacity operated by skilled crews. 'From the road [at Ouidah in Dahomey] where the ships

1 J. Bird, The Major Seaports of the United Kingdom, *pp. 27–35.*
2 H. P. White, '*The Ports of West Africa: Some Geographical Considerations*', Tijdschrift voor Economische en Sociale Geografie, *vol. 50 (1959) 1–8.*
3 J. Darbyshire, '*Sea Conditions at Tema Harbour*', Dock and Harbour Authority, *vol. 38 (1957) 277.*

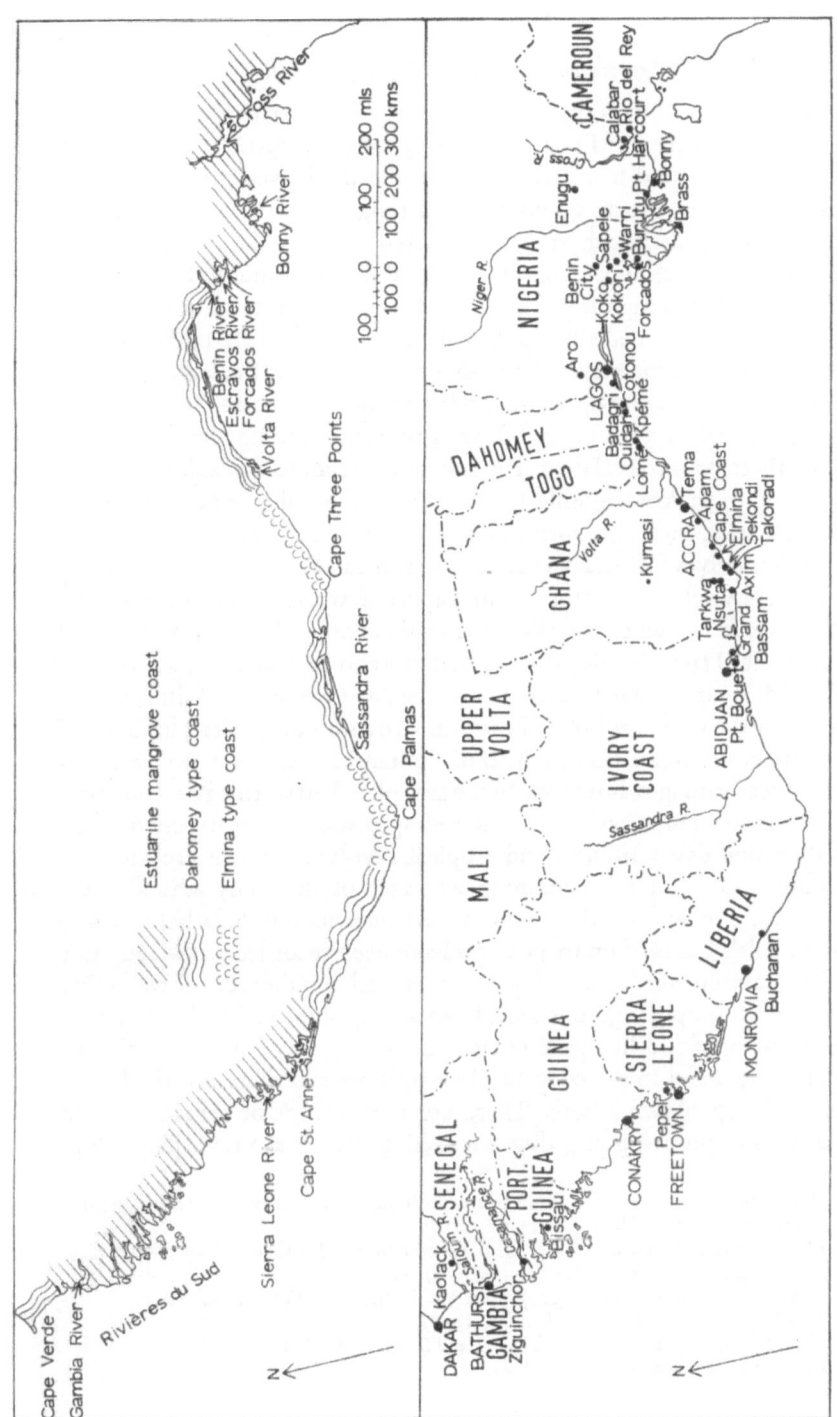

Fig. 2.1   West Africa: the nature of the coastline, and the pattern of port development

ride, there runs such a prodigious surf on shore, that a man risks drowning.'[1] A twentieth-century surf boat can take $1\frac{1}{2}$ tons of cargo and needs a crew of eleven.[2] Eastward from Cape Palmas, the south-westerly swell breaks at an angle to the beach, giving rise to a longshore drift, which has been estimated to move beach sand past a fixed point at the rate of 1 to $1\frac{1}{2}$ million tons per annum.[3] This creates an almost continuous offshore bar, behind which are longitudinal and transverse lagoons, the latter being the drowned valleys. Breaks in the bar are rare and the surf makes passage extremely dangerous.[4] Even if the sheltered waters of the lagoons can be reached, they are rarely deep enough for ocean-going ships. The rivers are generally small in volume and flow is restricted during the dry season. Even if there is no coastal dune, rivers fail to prevent formidable bars which may be up to $5\cdot6$ km. ($3\frac{1}{2}$ miles) wide.[5] Even where ships can occasionally enter estuaries, these are normally lined with mangrove swamps. These are in themselves formidable barriers, which can be penetrated neither by vessels nor by vehicles. Natural causeways are few and artificial ones costly.

In short, in the whole of West Africa there are only two good natural harbours, the Dakar Roads ('Rade de Dakar') and the Sierra Leone river, but both of these are at the western end of the coast (see Fig. 2.1). In addition, because of the obstacles listed above, the provision of artificial harbours is the more costly in relation to the traffic potential. There are two counter-balancing advantages, however. In the first place bad storms and dangerous seas are almost unknown. Winds exceeding Beaufort 7 are experienced on less than one day a month, and tropical revolving storms are virtually unknown. Only at the beginning and end of the rainy season is the coast subject to westward-moving disturbance lines, popularly known as 'line squalls'. Gusts of up to 45 knots have been recorded as the long line of black cumulo-nimbus clouds passes overhead, but these die down before they can whip up dangerous seas. Even sailing ships had little to fear, as the advancing cloud-line gave plenty of warning for shortening sail. The squalls are parallel to the coast or blow offshore and there is little danger of ships being blown ashore. There are few records of shipwreck, and losses among the sea-going dug-out fishing canoes are rare. Thus ships

---

1 R. *Astley*, A New General Collection of Voyages and Travels (*London, 1745*) II *412 (An account of Phillip's voyage in 1693)*.
2 *United Africa Company*, 'Surf Boat Operation in the Gold Coast', U.A.C. Statistical and Economic Review, *no. 5 (1950) 45*.
3 A. *Guilcher*, L'information géographique. I: Afrique (*Paris, 1954*) p. 57.
4 E. *Thorp*, Ladder of Bones (*London, 1967*).
5 *United Africa Company*, ' The Forcados and Escravos Bars', U.A.C. Statistical and Economic Review, *no. 5 (1950) 45*.

can lie offshore in safety at the surf ports, a great advantage in the early stages of development when the level of traffic was insufficient to justify capital investment.

The second advantage is the very limited tidal range. This is greatest between Cape Palmas and Cape Roxo (Portuguese Guinea). At Freetown it is only 3·2 m. (10½ ft) at Springs, at Bathurst to the north it is only 2 m. (6½ ft), east of Cape Palmas the amplitude rarely exceeds 1·5 m. (5 ft) and at Lagos it is only 0·9 m. (3 ft).[1] Vessels can therefore enter and leave port at any time, except when depth of water on a bar is really critical. They can also berth at any time and wet docks are unnecessary.

A second factor influencing port development has been the nature of the landward communications. Before 1897 land communications were primitive and costly. There were few waterways navigable by steamer or launch and otherwise reliance had to be placed on portering. This was about the most costly form of transport, even with the low wage rates pertaining at the turn of the century. In 1898 it took 1,400 carriers at a cost of £700 to move 31,500 kg. (70,000 lb.) of rubber from Kumasi to Cape Coast.[2] Under these circumstances, the need was for as many ports as possible along a given stretch of coastline. Because each had a very restricted hinterland and a very limited volume of trade, they received minimal investment in facilities. Before the First World War there were some thirty ports listed as open along the 530 km. (330 miles) of the Ghana coastline. However, any port which chanced to be near a navigable river leading to the interior would gain a competitive advantage over its neighbours. Thus Axim was located near the Ankobra river, which provided a relatively easy route for part of the way to the goldfield around Tarkwa, and could be used for floating down timber.[3] In the 1890s the mail boats called at Axim in preference to any of the half-dozen neighbouring ports. The early importance of St Louis du Sénégal and of the Niger Delta ports was due to their respective locations at the mouths of major rivers. Even the upgrading of a primitive track could have the same effect. It was reported that the 1875 military 'road' inland from Cape Coast was 'an immense improvement on the ordinary bush trails'. This can be connected with the fact that the neighbouring port of Anomabu was 'large but a shadow of its former self as trade now goes from Cape Coast'.[4]

1 *Admiralty*, The Africa Pilot (*H.M.S.O., 1953*).
2 *Gold Coast*, Annual Report, 1898, *Parliamentary Paper c.9498–5 (1900) para. 19.*
3 *Ibid.*
4 A. B. Ellis, West Africa Sketches (*London, 1881) pp. 50, 219.*

Between 1897 and 1911 railways were opened from seven ports – Conakry, Freetown, Abidjan, Sekondi, Lomé, Cotonou and Lagos[1] – which conferred competitive advantages on these ports. After the line from Tarkwa to Sekondi was opened in 1901, the mail boats began to call at the latter in preference to Axim. As soon as the line from St Louis to Dakar was opened in 1885, most of the trade on the Senegal river was transferred to Dakar, as it was much the better port. Again, the building of railways from Lagos and Port Harcourt to the north meant the concentration of the Nigerian trade on them to the detriment of the Delta ports, hitherto more important. The ports selected as rail terminals experienced a greatly increased volume of trade, and attracted ever more investment in facilities. They thus became the foci of the trunk-road systems which emerged during the 1920s and 1930s, making their competitive superiority even more marked.

Thus the development of internal communications tended to concentrate the expanding volume of trade on an increasingly small number of ports. Nowhere was this tendency seen better than in Ghana, where the number of ports open had fallen by 1965 from thirty to two. Conversely, in southern Liberia, where internal communications are sparse, thirteen small ports were reported as open in 1953 between Monrovia and Cape Palmas.[2] This tendency towards increasing concentration has been reinforced by the increasing cost of building and operating ships, which in turn has greatly increased the cost of keeping a ship idle in port. Shipowners, therefore, strive to reduce the number of calls per voyage and to cut down the time spent in each port, and there has been a demand for greater sophistication in cargo-handling methods at the ports. Increased investment must therefore be concentrated where it will produce the highest return – in the largest and busiest ports, where the equipment will achieve the highest utilisation. Almost 90 per cent of the West African trade is now concentrated at some ten ports.

To illustrate these tendencies, the coast can be divided into three major divisions:

1. *Cape Verde to Cape St Anne* (1,040 km. [650 miles]). Along this stretch submergence has drowned the lower basins of the rivers draining the Guinea Highlands, to create a series of large rias. Impressive as they may appear on the map, their utility is greatly reduced by the formidable

1 R. J. Harrison Church, 'The Evolution of Railways in French and British West Africa', Comptes Rendues du 16e Congrès International de Géographie (Lisbon, 1949) pp. 95–114.
2 Admiralty, The Africa Pilot.

estuarine bars and by the almost continuous fringe of mangrove swamps. Here is an interpenetration of land and sea rare in Africa, but the barrier of the surf has been replaced by a swamp barrier that is just as effective. One of the advantages enjoyed by Conakry and Freetown is that they are located on the ends of causeways, the Kaloum peninsula and the Colony Mountains respectively, jutting out through the swamps to deep water. These provide an approach for railways and roads. Bathurst, too, stands on hard ground upstream from the bar of the Gambia, which has 8·2 m. (27 ft) at low water. The river is navigable for a hundred miles for steamers drawing up to 4·6 m. (15 ft).

2. *Cape St Anne to the Benin river* (2,160 km. [1,350 miles]). Throughout this stretch the smooth, almost featureless coastline is continually beaten by the fierce surf. Here there are two contrasting types of coastline. The first is the dune and lagoon coast which may be termed the 'Dahomey type'. Along this coast dunes have eliminated the irregularities of the former coastline. Rivers can rarely breach the dunes, especially in the dry season, but when they do there is an impassable bar. There is only about 1·5 m. (5 ft) of water on the bars of the Sassandra, Ankobra and Volta rivers. The littoral dune is backed by marshes or lagoons. It is possible to travel by launch between Cotonou (Dahomey) and the Niger, some 440 km. (275 miles). But the lagoons are shallow and are deep enough to receive ocean vessels only at Abidjan and Lagos. This type of coast extends from Cape St Anne to southern Liberia, between the Sassandra river and Cape Three Points, and from the Volta to the Benin river.

The second type occurs where variation is provided by the close approach of low plateaux to the coast, and may be termed the 'Elmina type'. While there are dunes across the estuaries and valleys, usually backed by small lagoons, they are interrupted by a series of low bluffs, usually 6 to 15 m. (20 to 50 ft) high. These protect the beaches to the leeward or eastern side from the full force of the surf. The beach at Elmina in Ghana is the only one in the country on which an ordinary ship's boat can be landed, which was done during the Second World War.[1] In the author's view it was no coincidence that this site was selected in 1480 for the first permanent European settlement beyond Cape Palmas. There are two stretches of this type of coast: from southern Liberia to the Sassandra river, and from Cape Three Points to the Volta river.

1 *This information was given to the author in 1954 by the officer in charge of the Police Training Depot at Elmina Castle. It is supported by The* Africa Pilot: *'landing may be affected by ships' boats in Bera Lagoon at high water during the dry season'.*

B

3. *The Benin river to Mount Cameroun* (640 km. [400 miles]). In the neighbourhood of the Niger Delta the offshore contours shelve more gently and thus the swell breaks further out to sea. There are eighteen mouths of the Niger listed,[1] most of them 'dead', but all open to the sea. However, the eastward drift obstructs the entrances with bars that are constantly changing their profiles, necessitating continuous hydrographic survey. In 1922 the maximum advised draught for ships crossing the Forcado bar was 5·3 m. (17½ ft). By 1947 this had fallen to 4 m (13 ft). This passage was closed in 1940 and ships bound for the Delta ports were instructed to use the Escravos entrance with an advised draught of 4·3 m (14 ft). These branches of the Niger are lined with wide mangrove swamps, and dry approaches to the shores are very limited. Beyond the Delta itself, the estuaries of the Cross river and the Rio del Rey are essentially similar.

The pattern of port development

In response to the rudimentary nature of landward communications, West African coastal trade was centred from the fifteenth to the nineteenth centuries on numerous coastal 'factories', which were small ports with very restricted hinterlands. It would be difficult to estimate the exact number of ports open at a particular date, but during the latter part of the nineteenth century an average of about 150 were in operation on the 3,800 km. (2,400 miles) of coast. Although their equipment was uniformly simple, these ports were of two distinct types, surf and estuarine. The surf ports, which were completely open roadsteads, were the more common. If the stretch of coast was of the Dahomey type, the 'port' was generally no more than an open beach and its equipment a few surf boats. There were some simple buildings to house the factor and his stores, and a small village nearby.[2] A few ports, notably Badagri (Nigeria) and Ouidah (Dahomey), which remained the principal centres of the slave trade until the mid-nineteenth century, attracted numerous factories and there were a large number of permanent buildings.[3] The Elmina type of coast in Ghana provided sites for impressive castles, notably Elmina (1482), Cape Coast (1662), Fort James (Accra) (1673) and Christiansborg (1657). These were the result of international rivalry for the gold trade. All those between Axim and Accra, numbering about twenty-five, are on bluff

---

1 *Admiralty*, The Africa Pilot.
2 *R. Austen Freeman, the writer of detective fiction, was in the Gold Coast medical service in the 1890s. He provides us with some of the best impressions of these ports in his short stories; see 'Mutiny on the Speedwell' in* A Certain Dr Thorndyke.
3 *A good example is the John Holt store at Ouidah (Dahomey).*

sites, with protected beaches to leeward.[1] The western Ivory Coast, long
known as the 'côte des Mal Gens', had a less productive and actively
hostile hinterland, and there were no castles and few ports. The estuarine
ports were found in two main groups, those of the 'Rivières du Sud',
between Cape St Anne and the Saloum river, and those of the Niger
Delta and Cross river ('the Oil Rivers'). In the first-named area trading
stations grew up at the few possible landings that had land or water
communications with the interior. Here developed such ports as Kaolack
(Senegal), Bathurst, Bissau (Portuguese Guinea), Conakry and Freetown.
In all these cases roads can be built up to the port, and Bathurst is on the
navigable Gambia river.

With the extension of the hinterlands of a favoured few ports and with
a general increase in trade, improvements became justifiable. On the
Emlina type of coast it would be possible to supplement the natural
shelter of the bluff by a breakwater. As a result of a consultant's report in
1897, this was done at Accra to protect the surf-boat landing.[2] In 1911 a
breakwater was provided at Sekondi.[3] Similar works were proposed for
Elimna and Apam, but this form of improvement was not applied as
frequently as might be expected. The main limiting factor of the surf port
is the capacity of the boats (under 2 tons, and normally no single package
over 0·3 ton) together with the small number of round trips they can
make during a day's work. This could be overcome to some extent by
building a pier out through the surf. The swell beyond would be too great
to allow a ship to come alongside, but barges towed by tugs could transfer
cargo between pier and ships. Those used at Cotonou had a 15–20-ton
capacity, but 150-ton barges were used at Sekondi in the calmer waters of
Sekondi Bay.

Sekondi provided the sole example of this type of port in former
British West Africa. Within the shelter of the breakwater referred to, two
jetties with railway lines along them were built. The development of
manganese exports during the First World War was only made possible
because of this equipment, but at the same time it was this and other bulk
traffic which led to the construction of Takoradi harbour. The Germans
adopted the same method at Lomé in 1904 and the French at Cotonou in
1892. It was also used at Port Bouet and Grand Bassam in Ivory Coast. It
will be noted that all these ports, except Sekondi, were on coasts of the

---

1 *None of the fairly numerous studies of Gold Coast castles has drawn attention to this
   fact.*
2 *Coode, Son and Matthews*, Report on Gold Coast Harbours, *7 Aug 1897.*
3 *Gold Coast, Sessional Paper vi (1918–19).*

Dahomey type. The completion of the Sekondi works allowed the closure between 1919 and 1921 of six neighbouring surf ports, except for some timber through Axim. The completion of the wharves at Cotonou and Lomé led to the abandonment of all other surf ports in Dahomey and Togo.

So far no development provided complete protection from the swell or allowed ships to berth alongside properly equipped quays, the only way of radically increasing port capacity and reducing ship turnround times. The artificial harbour at Dakar was begun in 1863,[1] but Dakar is in many ways exceptional. It has close ties with the main shipping routes of the South Atlantic; at an early date it became a permanent base for the French Navy, and it remained the only example of an artificial harbour in West Africa for many years.

In 1920 exports through Sekondi reached 97,205 and imports 84,434 tons, and both were rising fast. In 1918 manganese was available at the Nsuta mine at 20s per ton, but reached 36s 4d per ton f.o.b. at Sekondi, 48 km. (30 miles) away – an 82 per cent increase. These facts and the limitations of the other Gold Coast surf ports led the Government to plan an artificial harbour at Takoradi,[2] opened in 1928. A southern breakwater, 1,280 m. (1,400 yds) long, protects a large water area, enclosed to the north by a wide mole, at which vessels can berth. After 1945, increasing traffic made further large-scale investment imperative, and extensions went into operation in 1955. Takoradi now has four general cargo berths and a manganese berth on the southern side of the mole, where there is 9–11 m. (30–36 ft) of water, and oil and bauxite berths on the north side. In addition there are eight moorings for loading timber. In 1952 work started on another artificial harbour at Tema, 24 km. (15 miles) east of Accra. Here three miles of moles have been built to enclose 170 hectares (430 acres) of water. There is a five-berth finger quay springing from the western mole and a seven-berth marginal quay. There is also an oil berth and a fishing harbour. The port came into full operation in 1962.[3]

The construction of Takoradi and Tema also entailed the building of road and rail links with the national systems, and of two new towns each with a population of over 50,000. Altogether they represent a vast capital investment; and they have brought to a logical conclusion the process of port consolidation which began with the development of Sekondi. In

1 *Naval Intelligence Division*, Geographical Handbook, BR 512: French West Africa, *vol. 1*.
2 *Gold Coast, Sessional Paper viii (1920)*.
3 D. Hilling, '*Tema, the Geography of a New Port*', Geography, *vol. 51 (1966)* 111–25.

1914 there were thirty surf ports open in Ghana, but by 1930, with the opening of Takoradi, these had been reduced to nine; after the Second World War, with better roads to Takoradi and Accra, there were only five officially open, and by 1962 there were only two.[1]

During the Second World War the Americans built a similar artificial harbour at Monrovia in Liberia. Another was planned by the French to serve Togo and Dahomey, but inter-territorial rivalry led to separate developments at Lomé and Cotonou. A feasibility study recommended in 1956 that a proposed artificial harbour scheme at Cotonou was technically and economically justifiable, particularly in view of the fact that most of Niger's overseas trade passed through this port. Cost reduction was estimated at £2 per ton and this appeared to justify the £10 million investment. Work started in 1959 and in 1965 the new port was opened.[2]

A similar harbour has been built, with less economic justification, at Lomé in Togo. When current plans are completed there will be seven Takoradi-type ports along the coast, superseding dozens of the original surf ports. These will include Takoradi and Tema in Ghana, Lomé (Togo), Cotonou (Dahomey), Monrovia and Buchanan (Liberia), and San Pedro (Ivory Coast) on which work has just commenced. Dakar, with ample natural protection from the tombolo of Cape Verde, lacks the long moles and is not of this type, but here again development had led to the decline of St Louis and of four surf ports between Dakar and the Saloum estuary. Of the three estuarine ports of the Saloum only Kaolack survives, with a diminishing trade, and a small amount of traffic is still handled at Ziguinchor on the Casamance. With these exceptions all Senegalese trade passes through Dakar.

The development of Lagos and Abidjan has followed a different pattern, for here the lagoons are deep enough to receive ocean vessels. The problem has been to provide an adequate entrance through the littoral dunes, as Lagos is the only natural entrance to the lagoons systems in the 328 km. (205 miles) between the Benin river and Cotonou, but the bar was shallow and dangerous. Early records show a variation of between 2·7 and 4 m. (9 and 13 ft),[3] and in six months during 1891 the 'bad bar' flag forbidding the crossing was flown on fifty-six days. Passengers and mail were transferred to 'branch steamers' in the roads, but most cargo had to

1 *Gold Coast*, Annual Reports.
2 D. Hilling, '*The new Port of Cotonou*', Fairplay, *7 Oct 1965, Cargo-handling supplement, p. xiii.*
3 *Despatch by Clifford to Secretary of State, 12 Nov 1919 (Nigeria 1016).*

be transhipped at Forcados.[1] Numerous improvement schemes were proposed, but it was finally decided to interrupt the eastward drift by moles, which would allow the bar to be dredged. The east mole was begun in 1907, using stone railed 93 km. (58 miles) from Aro, and the west mole in 1910. The first mail boat entered the lagoon in 1914 and by 1920 there was 6 m. (20 ft) of water on the bar.[2] The present depth is 8·2 m. (27 ft).

Difficulties have arisen from the accumulation of sand on the windward or Lighthouse Beach, which threatens to reach the end of the mole before the century is out, and from the concomitant erosion of Victoria Beach to the east. Already storms have created temporary breaches, in spite of the deposition on this beach of material dredged from the harbour.[3] The only ultimate solution is the pumping of sand from west to east under the entrance. Within the harbour, three berths were provided by 1911 on Lagos Island at Customs Wharf. In 1920 it was decided to build a four-berth quay on the mainland at Apapa, so that road and rail links would avoid the congested Lagos Island. This quay has been gradually extended by reclamation of the lagoon and there are now thirteen berths for general cargo and two for petroleum, with ample room for extensions on Badagri Creek.

Development at Abidjan followed very similar lines, but here an artificial entrance was needed. An attempt was made in 1904–7 but the channel was soon closed by the coastal drift. Construction of the Vridi Canal was begun in 1937, but because of the war ships did not enter the lagoon until 1950, when Abidjan finally superseded the pier at Port Bouet. The 2·8 km. (1¾ mile) canal is aligned at such an angle to the coast and has its entrance so shaped that the scouring action of the tide is maximised, but this has led to erosion of the leeward beach. Within the harbour deep-water berths were provided on the Île de Petit Bassam. Six were completed by 1955, but more have been provided since, for traffic doubled between 1950 and 1956.[4]

There is no such clear pattern discernible among the estuarine ports, but it can be said that over the years increasing trade has led, where possible, to the provision of deep-water berths and to improvements to the approach channels. This has led to concentration of investment and subsequent concentration of traffic on a few favoured ports. The Rivières

1 *Unpublished report by Coode, Son and Matthews, 1892.*
2 *D. C. Coode, 'Design and Construction of Apapa Wharf Extension'*, Proc. Inst. Civ. Eng., N.S., *vol. 7 (1957) 499–536.*
3 *J. E. Webb*, The Erosion of Victoria Beach (*Ibadan, 1960*).
4 *Sharlet, op. cit.*

du Sud have seen the usual process of consolidation, and there now remain only two ports of any importance, Freetown and Conakry. Cargo was lightered ashore at Freetown until the completion of Queen Elizabeth II Quay in 1954. This has two deep-water berths and there is thus, because of the restricted hinterland, a vast disparity between the port of Freetown and the natural harbour of the Sierra Leone river, which has anchorage for 150 vessels of 'unrestricted draught'. This is sheltered by a wide bar, but the tides keep a deep channel scoured at the southern end.[1] Conakry, on Tombo Island at the end of the long Kaloum peninsula, might seem oddly classified as estuarine. But it has many characteristics of other Rivières du Sud ports, including the dry land approach and freedom from surf. A deep-water quay is provided on the north of Tombo Island, which offers protection from the southerly swell, while an island mole shelters it from the north. There are two general berths and specialised ones for iron ore and bananas. Bathurst, Ziguinchor and Bissau are of minor importance and owe their continued existence largely to the alignment of the international boundaries.

The once numerous 'Oil Rivers' ports have suffered the familiar process of consolidation, and once busy places such as Akassa and Brass are now dead. The survivors are in three groups: those reached by the Escravos and Bonny entrances, together with Calabar on the Cross estuary. The first group consists of Sapele, Warri, Burutu and Koko, on various arms of the Niger, but accessible from the sea only across the Escravos bar. Depth of water here controls the size of vessels entering and therefore the capacity and operating costs of the ports. To improve the bar a five-mile mole has been constructed of stone brought from Ore along 48 km. (30 miles) of railway built for the purpose and thence 145 km. (90 miles) by lighter. Dredging has already produced 7·6 m. (25 ft) on the bar, but it is doubtful whether the investment will ever be justified because of limitations of the landward communications.

Sapele and Warri can be reached by road from Benin City and have become the chief outlets of the Mid-West State, but the resources of the region are limited except for the oilfields now being developed.

Koko is much nearer Benin, but although a port here has long been advocated, the recently completed £1 million deep-water quay is unused, and no ships use the port. Cargo is lightered at Sapele for the most part, though the African Timber and Plywood factory has a private deep-water wharf. At Warri there is one berth, but two more are under

1 H. R. Jarrett, 'Recent Port and Harbour Developments at Freetown', Scottish Geographical Magazine, vol. 71 (1955) 157–64.

construction to deal with materials imported for developing the oilfields. Burutu, with two berths, has no road access and, with Warri, is the terminus of the Niger fleets. The tonnage moved inland by water has remained almost unchanged for many years. Successive consultants' reports have advocated increasing river traffic, but road and rail developments tend to preclude this. The author is of the firm opinion that the investment in the Escravos project would have been better spent in improving the approaches to and facilities at Lagos and Port Harcourt.

The Bonny entrance gives access to Port Harcourt, discovered more or less by accident in 1915 by a surveying party charged with finding a terminal for the railway from the Enugu coalfield. Subsequent road and rail developments have extended its hinterland as far as Bornu and Kano and it has become the second port of Nigeria. Berths have been constructed on an inside bend of the Bonny and have been extended several times. The development of the oilfield and the building of a refinery increased trade in recent years, but the Nigerian civil war produced highly adverse effects. Calabar grew to importance as a centre for river traffic. With the transfer of trade to road traffic, the port suffered in competition with Port Harcourt, since until recently there were ferries on all the road approaches. Development of the almost virgin East Cross river basin may increase its trade again.

Continuing economic development in West Africa has led to the latest stage in morphological development, the emergence of the specialised port. Mention has been made of specialised berths in general ports, but the first specialised port, though legally part of Freetown, was that of Pepel on the Sierra Leone river. Pepel, 24·8 km. (15½ miles) upstream from Freetown, was developed in the 1930s as the terminus of a private railway from the iron mine at Lunsar. It has a long pier into deep water. Loading is fully automated, about 2 million tons of ore a year being shipped. Another specialised iron-ore port has been developed at Buchanan in Liberia. At Kpémé, 35 km. (22 miles) east of Lomé, a pier was opened in 1961 to facilitate phosphate exports, which reached 1·1 million tons in 1966. The pier, used to support a conveyor belt from the processing plant, is 1,180 m. (1,300 yds) long and there is 12·9 m. (42½ ft) of water at the outer end. Ships are moored to buoys independently of the pier as the swell is considerable. This represents the final stage of the 'Cotonou type' of port.

In the Niger Delta several petroleum ports have been developed. There has even been a reversion to hulks, though the modern ones consist of the bow portions of two old tankers joined together. One such is moored

18·4 km. (11½ miles) off the Escravos, where the oilfield is a submarine one. Two and a half million tons were shipped in 1966. Other similar terminals are being developed off the Brass and Pennington rivers. But the principal facilities are at Bonny town, inside the Bonny bar, which has been dredged from 7 to 11·3 m. (23 to 37 ft), the deepening of the channel costing £3 million. There are six moorings connected to tank farms, themselves connected with the oilfield and with the Port Harcourt refinery. Exports grew from 1 million tons in 1960 to 20 million in 1966. The civil war which broke out in 1967 has unfortunately brought trade over Bonny bar to a complete standstill at the time of writing.[1]

In this essay an attempt has been made to analyse the stages by which the primitive surf and estuarine ports along the West African coast have evolved into well-equipped and up-to-date ports, which lack nothing by world standards. This process has been accomplished in well-defined stages and a distinct pattern is discernible. The nature of the coastline has influenced the morphological development of the 'seaward' side of the port, and competitive expansion of port hinterlands as the result of road, rail and river improvements has influenced that of the 'landward' aspects. In general, however, the process began with a large number of small ports and has involved the concentration of the ever-growing trade into fewer, larger and better-equipped ports. The distribution of these terminals depends to a large extent upon the location of international boundaries. Each West African country, with the exception of Portuguese Guinea, is already equipped with, or is currently planning to develop, one or more large modernised ports, with the aid of artificial shelter where necessary.

1 *In all cases details of specialised ports were obtained during field work.*

3. Nouadhibou (Port Étienne) and the Economic Development of Mauritania

Charles Toupet

Dr Charles Toupet is Maître-Assistant in the Department of Geography in the Faculté des Lettres et Sciences Humaines of the University of Dakar, Senegal. His research interests and publications lie mainly in the fields of biogeography and the human and economic geography of the Sahel zone of West Africa, with special reference to Mauritania.

The traditional economy of Mauritania is based essentially upon nomadic pastoralism, oasis cultivation and caravan trading, and most Mauritanians feel a strong disregard for the sea as a source of economic gain. Nouadhibou[1] has thus appeared for long as an insignificant point of reference, a drop of water lost in a continual haze of sand, isolated on the boundary between the immensity of the desert and that of the ocean. The Mauritanian writer Ahmed Lamine ech-Chenguiti (1911)[2] makes no mention of Port Étienne in his account of the geography of the country; and along the low inhospitable stretch of coast between Lévrier Bay and Nouakchott the only inhabitants are a few hundred fishermen of the Imraguen tribe,[3] half-caste descendants of the Berbers and their Negro slaves. Nouadhibou is in fact separated from the core area of Mauritania by considerable distances and by natural barriers to communication which include the active dune areas of the Akchar and the Azeffal.[4] The Sahel zone, south of the 100-mm. (3·9-in.) isohyet (see Fig. 3.1), contains 90 per cent of the population of the country and is essentially a stock-rearing area orientated towards Senegal and its great port of Dakar.[5] Three principal factors largely explain the development in this unpromising situation of a new port and town scheduled to take its place amongst the more important

1 *The town's name was officially changed from Port Étienne to Nouadhibou in January 1969.*
2 *A. L. Chenguiti,* El Wasit: Littérature, histoire, géographie, mœurs et coutumes des habitants de la Mauritanie *(St Louis du Sénégal, 1953).*
3 *Lt Revol, '*Étude sur les fractions d'Imraguen de la côte de Mauritanie*', Bulletin du Comité d'Études Historiques et Scientifiques de l'A.O.F., vol. 20 (1937) 179–224.*
4 *C. Toupet, '*Le problème des transports en Mauritanie*', Bulletin I.F.A.N., série B, vol. 25 (1963) 80–106.*
5 *Ibid.*

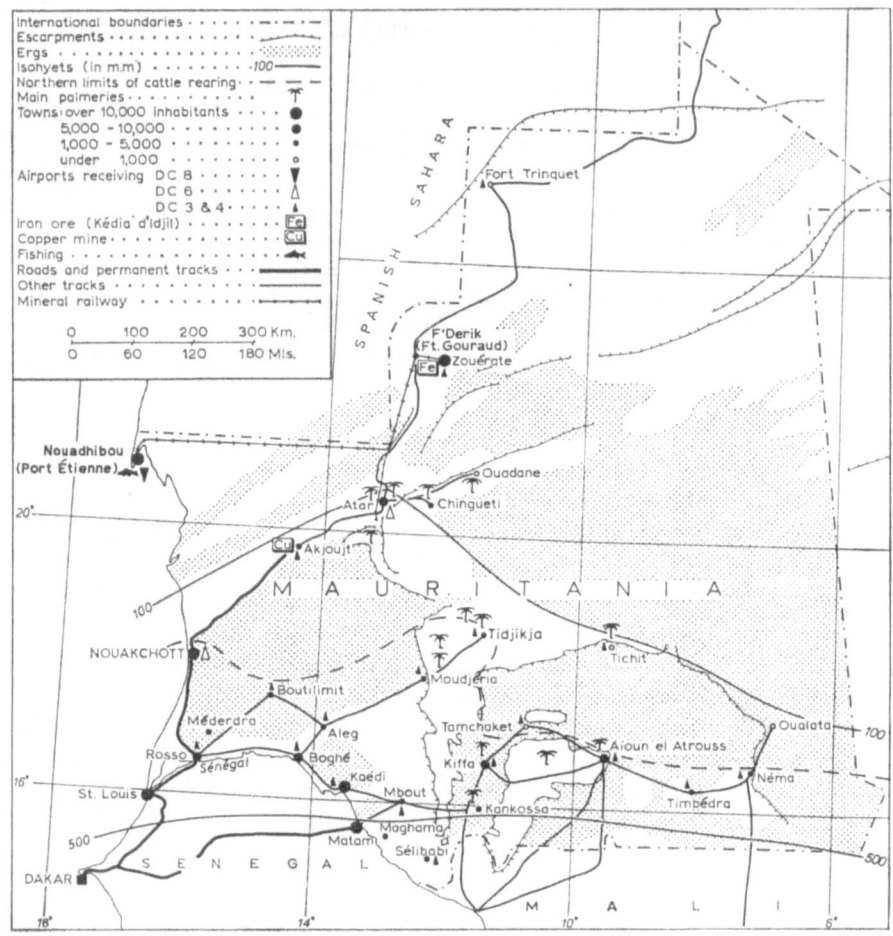

Fig. 3.1   *Economic map of Mauritania*

trade centres of tropical Africa: site conditions, the fishing industry, and the exploitation of iron-ore resources.

The Cap Blanc peninsula (Fig. 3.2) stretches for 50 km. (30 miles) from north to south and encloses an extensive bay, the origin of which was probably associated with a fault.[1] Lévrier Bay offers two principal advantages. Firstly, unlike Rio de Oro (Spanish Sahara), it offers depths exceeding 10 m. (33 ft) at Nouadhibou and 15 m. (50 ft) at Cap Blanc,

1 A. Blanchot, '*Les formations récentres de Mauritanie occidentale*', Bulletin de la Direction Fédérale des Mines et de la Géologie, *vol. 20 (Dakar, 1957) 9–93.*

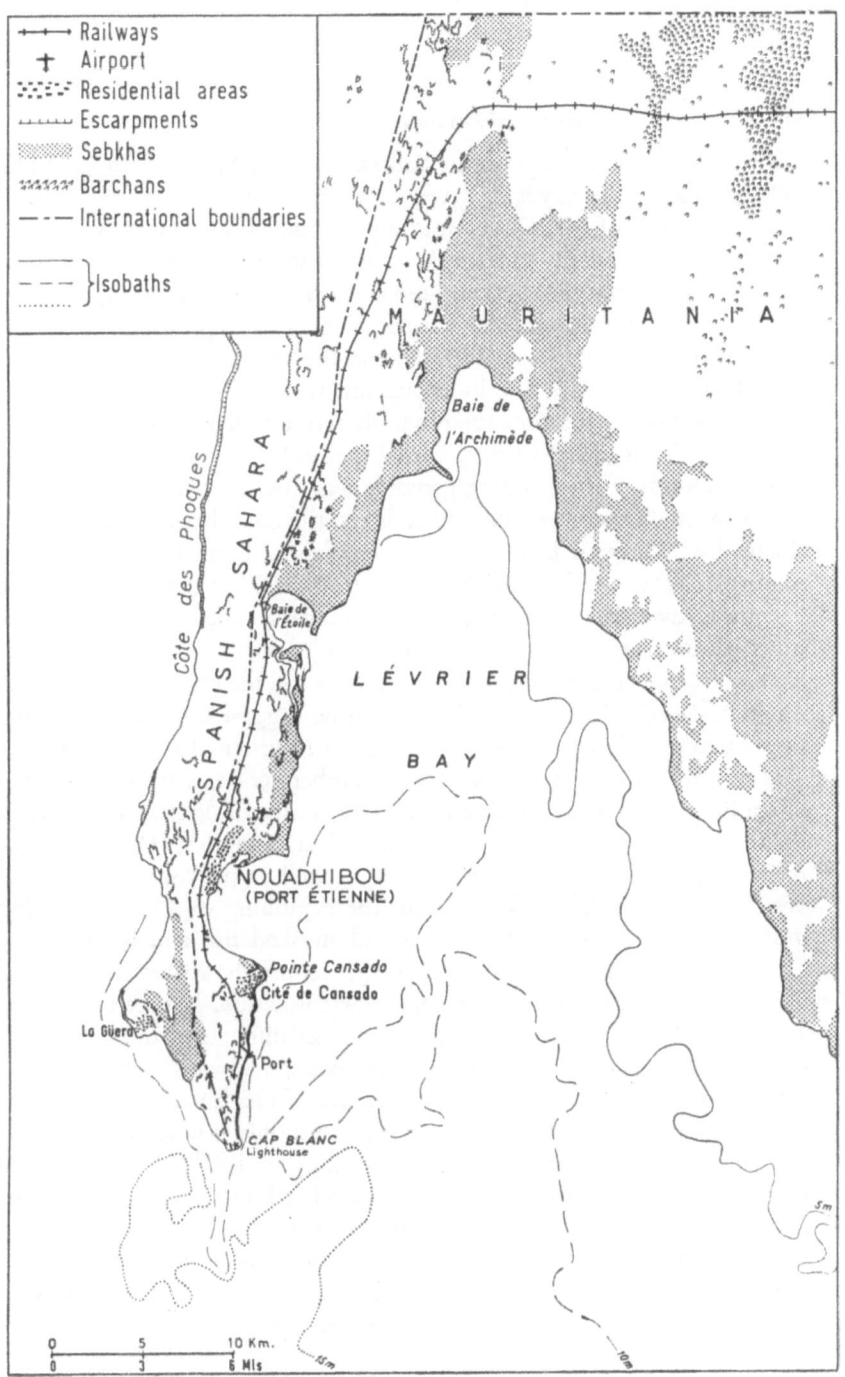

Legend:

- +→ Railways
- ✝ Airport
- Residential areas
- Escarpments
- Sebkhas
- Barchans
- – · – · International boundaries
- – – } Isobaths

MAURITANIA

Baie de l'Archimède

Côte des Phoques

SPANISH SAHARA

Baie de l'Étoile

LÉVRIER

BAY

NOUADHIBOU
(PORT ÉTIENNE)

Pointe Cansado
Cité de Cansado

La Güera

Port

CAP BLANC
Lighthouse

5 m

0   5   10 Km.
0   3   6 Mls

Fig. 3.2   The site of Nouadhibou (Port Étienne)

and so is the only harbour on the Saharan coast of West Africa accessible to large vessels. The bay is sheltered from the strong Atlantic swell which attacks the cliffs of the 'Côte des Phoques', and its calm waters are only occasionally ruffled by an easterly wind. Moreover, the tidal range is relatively slight, varying between 0·6 and 1·8 m. (2–6 ft) during the year. Secondly, the waters off the Mauritanian coast represent one of the world's richest fishing grounds. Along the coast, cold water rich in phosphates and nitrates rises in association with the Canary Current and provides abundant food for the phytoplankton, which in turn encourage enormous shoals of fish of very varied species.[1]

The richness of these fishing grounds was known to Portuguese navigators in the fifteenth century, since when the coast has been frequented chiefly by fishermen from the Canary Islands. Early in the twentieth century, following the French penetration of Mauritania, the French scientist Gruvel undertook in the course of several expeditions a systematic study of the marine fauna of the region and discussed the possibilities of establishing a fishing industry.[2] Gruvel proposed the establishment on the shores of Lévrier Bay of a station which he suggested should be named Port Étienne in honour of the then French Minister of Colonial Affairs. The period between May 1906 and September 1908 saw the construction of a military fort, a residency, a dispensary, a meteorological station and a lighthouse at Cap Blanc with a range of 29 km. (18 miles). These developments provided a measure of security as well as provisions for fishing vessels, and were designed to permit the beginning of a fish-processing industry, based initially upon dried and smoked fish and later upon a fish-canning plant.[3] Various problems prevented the realisation of these schemes, however, until very recent years: isolation, lack of population and scarcity of drinking water were the chief difficulties, the last-named due not only to low rainfall (the annual average between 1931 and 1960 was 31 mm. [1·2 in.]) but also to the presence of a highly saliferous crust in the subsoil below which only a very meagre supply is obtainable. Only a few small private enterprises survived, dealing in dried fish and operated largely by fishermen from the Canary Islands who took advantage of the low costs involved and of the climatic conditions which are highly favourable for fish-drying: the average temperature in the coldest month is 19°C. (66·2°F.) and in the warmest month 25°C. (77°F.); the dry trade

1 A. Guilcher, Précis d'hydrologie marine et continentale (*Paris, 1965*).
2 A. Gruvel, and R. Chudeau, À travers la Mauritanie occidentale: de Saint-Louis à Port-Étienne, 2 vols (*Paris, 1909–11*).
3 A. Gruvel, L'industrie des pêches sur la côte occidentale d'Afrique, du Cap Blanc au Cap de Bonne Espérance (*Paris, 1913*).

winds are constant, and the diurnal variation in relative humidity is between 83 per cent at 6 a.m. and 57 per cent at noon.

An external stimulus was obviously needed to stir Nouadhibou out of its lethargic state, and this came eventually in the form of the discovery and exploitation of the iron-ore resources of Kédia d'Idjil, near F'Derik (Fort Gouraud) (Fig 3.1), estimated at 200 million tons with an average iron content of 64 per cent. Between April 1960 and June 1963, when the installations of the Société des Mines de Fer de Mauritanie (MIFERMA) were opened, the peaceful straggling village of Nouadhibou was transformed into an immense building site where European engineers, Mauritanian officials and Senegalese and Mauritanian employees worked side by side. It was necessary to install at Nouadhibou all the facilities and equipment needed for the unloading of thousands of tons of cement, machinery and railway materials required for the development of the mining centre at Zouérate and of a 650-km. (400-mile) railway link to the coast. The project also involved the construction, 10 km. (6 miles) south of the town, of an ore-stockpiling zone and a mineral port in deep water capable of receiving and loading with minimum delay the largest modern ore-carriers.

The MIFERMA company has successfully overcome the problems of water supply and lack of population, and has broken the traditional isolation of Nouadhibou by building the railway which serves the greater part of northern Mauritania, by developing the commercial port and by greatly enlarging the airport in order to receive four-engined jets (Table 3.1). Fresh water supplies, formerly brought in by water boats, now come from a 120-m. (395-ft) borehole at Bou Lanouar, 90 km. (56 miles) east of Nouadhibou, by means of tanker wagons attached to iron-ore trains. This procedure proved expensive and from 1969 a more economical piped water system has been used.

These developments have attracted a considerable population of workers and their families to the port. The total population of Nouadhibou is today about 12,000; the port is thus second only to the capital, Nouakchott, in terms of size. Most of the inhabitants are Mauritanian citizens: business men and civil servants, employees of MIFERMA and numerous unemployed persons attracted out of the wilderness by the activities of the expanding town. Most of the Senegalese employed in the construction work have now returned home. The French element in the population includes several hundred industrialists, engineers and technicians, and civil servants administering technical aid, together with their families.

Table 3.1

Air traffic at Nouadhibou, 1960–7

	Aircraft Movements*	Passenger Movements†	Freight Movements (tons)
1960	538‡	5,081	170
1961	773	17,075	571
1962	768	17,994	774
1963	1,067	22,679	541
1964	1,560	17,942	570
1965	1,443	22,460	650
1966	1,900	28,362	837
1967	1,550	36,429	802

* Arrivals and departures.

† The total figures given include arrivals, departures, and passengers in transit.

‡ 11 months only.

An interesting contrast has developed between the two chief foci of the Nouadhibou settlement. The town itself, spreading around the Baie du Repos, offers a variety of architectural indications of its chequered history: old battlemented houses, whitewashed Canary cottages, modern warehouses grouped around the port, and everywhere the fish-drying installations, with not a single tree to soften the harsh outlines of a stone town often half-hidden in a haze of sand. The new mining town built by MIFERMA on the Cansado plateau offers a different prospect. Here, sheltered from the sand, the company has developed in an hierarchical pattern some thirty villas for senior executives, 170 living units for senior staff and 490 living units for junior staff.[1] Modern amenities include a dispensary, a school, sports facilities, a club, a cinema and a department store. The small Spanish port of La Güera forms a third element in the Nouadhibou complex. Founded in 1920 a short distance within the Spanish Sahara boundary, the port lies on a small creek which indents the cliffs at this point and can receive only small fishing or surf boats. The

1 V. Marbeau, 'Les mines de fer de Mauritanie: MIFERMA', Annales de Géographie, vol. 74 (1965) 175–93.

settlement has remained small in spite of the recent development of fish-drying factories, and consists chiefly of an old fort, a lighthouse, a tiny church, some whitewashed cottages and a few shops offering tax-free bargains to visitors from Nouadhibou.

The fishing port

The MIFERMA organisation has provided a vital stimulus to the development of the three distinct spatial units within the Nouadhibou complex and to the expansion of three spheres of economic activity: the fishing industry, commercial development and the export trade. The decisive advantage of Nouadhibou as a fishing centre is the proximity of the rich fishing grounds which extend over the continental shelf some 145 km (90 miles) offshore as far south as latitude 20° N. The variety of both local and migratory fish species is so great that fishing takes place all the year round. Five main types of fishing may be distinguished: deep-sea trawling, tunny fishing, lobster catching, seasonal fishing from January to July for certain types of small local fish, and line- and net-fishing from the shore as practised by the Imraguen.

Trawling has increased in importance considerably during the last ten years. To the Spanish, Portuguese, Greek, Breton and Basque trawlers frequenting these grounds are now added fleets of modern Japanese, Soviet, Polish and Israeli vessels. The annual total catch is now estimated at about 200,000 tons, but the tonnages of tunny and lobster are not precisely known. Apart from a few fishermen who take advantage of the opportunity to send fresh lobster and other expensive fish to European markets by air, most of the trawlermen and the lobster and tunny fishers do not put in at Nouadhibou. The two fish-drying establishments in the port – the Société Industrielle de la Grande Pêche (S.I.G.P., founded in 1919) and the Entreprise Générale Atlantique (E.G.A., founded in 1957) – are supplied exclusively by fishermen from the Canary Islands who land only a few thousand tons a year. The fish is salted, cleaned and dried on wooden trays on the ground or suspended on a trellis; these simple operations in very favourable climatic conditions allow low-cost production of high quality. In spite of competition from the Canary Islands and from Angola, Mauritania exports an average of 4,000 tons of dried fish a year, mainly to Gabon, Congo (Brazzaville) and Congo (Kinshasa). The Imraguen fishermen handle only small quantities of dried fish and *poutargue* (mullet eggs, the Mauritanian equivalent of caviare). The further development of Nouadhibou could readily include a local fishing fleet based upon modern, well-equipped facilities in the port itself.

The Société Mauritanienne de Pêche (SOMAP) developed in 1966–8 a fleet of two 52-m. (170-ft) trawlers and six 32-m. (105-ft) trawlers for dragnet fishing together with six other vessels. This company is endeavouring to build up Mauritanian crews under Breton captains, and is obtaining encouraging results in spite of the traditional Mauritanian ignorance of, and revulsion for, the sea. With the aid of a loan from the Fonds Européen de Développement (F.E.D.), of 1,270 million francs CFA, Nouadhibou is now well equipped as a fishing port. Facilities include a fishing wharf 285 m. (310 yds) long, with a minimum depth alongside of 6 m. (20 ft); a cold store for the conservation of fish, and a covered fish market 108 m. (118 yds) long. These facilities are supplemented by three other large-scale enterprises: a refrigerated warehouse built in 1966 with private capital, which has an annual capacity of 20,000 tons of frozen fish and fillets (7,000 tons were handled in 1967); a factory for the manufacture of fish flour, which can treat 600 tons of fresh fish daily, and a group of installations developed by the IMAPEC company with Spanish finance comprising a fish-drying factory designed to treat 20,000 tons of fresh fish annually, a fish-meal factory treating 100 tons of fresh fish daily, and a fish-canning factory with an annual capacity of 3,000 tons. When these developments are fully operational, in 1969–70, Nouadhibou will be capable of handling some 140,000 tons of fresh fish a year, a figure which raises considerable problems of overfishing as well as problems of market outlets.

In the absence of any detailed biological study of the conditions of reproduction, feeding and migration of the various fish species found off the coast of Mauritania, precise indications concerning the size of shoals and the possibilities of overfishing are impossible; but many authorities now take the view that serious overfishing may already be taking place. Quantities of fish landed are regularly underestimated, biotopes are destroyed by the careless activities of trawlers, and during the 1967–8 winter season the tonnages of fish landed showed a considerable decline. Various steps towards conservation have been taken by the Mauritanian Government. Trawling is absolutely forbidden within Lévrier Bay, and within a 10-km. (6-mile) territorial limit is only permitted to Mauritanian nationals and to fishermen from those countries which have made specific arrangements with Mauritania, notably France and Spain. Fishing in the contiguous zone between the 10-km. (6-mile) and 20-km. (12-mile) limits is subject to application for permission. Although Mauritania has repeatedly denounced unauthorised fishing within the 20-km. (12-mile) limit, the problem is a growing one since adequate supervision of the controlled zone cannot yet be exercised and no international agreements cover the depths beyond these limits.

The problem of marketing is equally grave, for although the development of dried-fish exports to tropical African states presents few problems, the growth of output of frozen and canned fish and of fish-meal demands accurate market surveys and investigation of pricing problems, and as yet these have not been made in adequate depth.

The commercial port

In order to handle efficiently the traffic engendered by the MIFERMA developments, various facilities were completed at Nouadhibou in 1960–1. A wharf 80 m. (87 yds) long and 25 m. (27 yds) wide, with a depth alongside of 8 m. (26 ft), can receive most vessels frequenting the West African coast. The wharf is linked by a jetty 225 m. (245 yds) long to a lighterage quay behind which are located various transit sheds and offices. All port operations are controlled by the Société Mauritanienne d'Accorage et de Manutention (SAMMA), a subsidiary of MIFERMA; facilities include eleven cranes, three hoists, five tractors, six tugs, fourteen lighters and a 120-ton floating crane. Transit sheds and warehouses are provided with a total area of 3,000 sq. m. (32,280 sq. ft).

Harbour dues and charges are relatively low at Nouadhibou, and include only a simple berthing tax and a handling charge varying according to the nature of the cargo. Customs tariffs are identical with those in force at Dakar. Five shipping lines serve the port regularly: Maurel et Prom (Bordeaux) every twenty-five days; Paquet (Marseilles) once a month; the Danish Dafra Line three times every two months; the Spanish Aucona Line fortnightly; and the S.N.I.E. Line from Dakar monthly. Various other vessels call occasionally, most belonging to the Ghanaian Black Star Line or to French companies such as the Chargeurs Réunis and Dalmas-Vieljeux; British vessels of the Elder-Dempster Line call from time to time, as do some German vessels.

In the absence of reliable customs records it is unfortunately impossible to give precise details of shipping movements at Nouadhibou. In broad terms it appears that the number of ships entering the port in 1959 was approximately twenty, with a net registered tonnage of 50,000; by 1962 these figures had risen to 260 and 500,000 respectively, but by 1965 had fallen off to 210 and 350,000. In contrast, cargo traffic movements are well documented by SAMMA (Table 3.2). The main characteristics are the predominance of imports over exports – overwhelming in certain years – and the wide variation in the volume of imported goods as a result of the high proportion of construction materials, demand for which fluctuates greatly. Import figures reflect particularly the development of the mining complex; in 1962, out of a total volume of 168,076 tons of cargo imported, MIFERMA materials accounted for 153,746 tons (91 per cent, of which 86,176 tons were rails and sleepers). Subsequently imports have decreased appreciably, generally totalling about 40,000 tons and

Table 3.2

Cargo traffic handled at Nouadhibou, 1961–8

	Imports		Exports		Total
	MIFERMA imports	Total imports	Dried fish exports	Total exports	
1961	—	148,430	—	5,020	153,450
1962	153,746	168,076	—	5,988	174,064
1963	56,508	69,204	3,381	9,793	78,997
1964	19,388	31,380	3,456	6,946	38,226
1965	30,010	38,927	4,438	8,639	47,566
1966	18,884	36,876	4,652	6,321	43,197
1967	22,504	47,366	4,305	11,770	59,136
1968	15,134	43,763	—	14,908	58,671

Source: *Société Mauritanienne d'Accorage et de Manutention.*

made up chiefly of cement, scrap metal, machinery and materials for the industrial and port installations in process of development. The MIFERMA share of the total volume of cargo imported in 1968 was 34 per cent. As for consumer goods, their considerable diversity hardly serves to mask their comparatively insignificant volume; the chief items are usually flour, rice and sugar (about 1,000 tons each), together with smaller quantities of tea, hardware, medical supplies, etc.

Exports through the commercial port are extremely limited. The most important commodity is dried fish. Since 1964, however, exports of scrap iron have been of some significance (2,651 tons in 1965) and may rise considerably in the future. The broad outlook for the traffic of the commercial port in the immediate future is not very encouraging. The completion of the main port and industrial construction work in 1969 suggests that imports of building materials and equipment are likely to be much lower in future years. Imports of consumer goods will increase only in proportion to increases in employment in the port and its industries, which in their turn should form an expanding source of exports.

The mineral port

The site of the mineral port at Point Central is particularly favourable. Situated some 10 km. (6 miles) south of the town, the port is located on a sheltered stretch of water, backed by sandstone cliffs, where depths exceed 15 m. (50 ft) at a distance of 450 m. (490 yds) from the shore. A link with the open sea is provided by a natural channel with a minimum depth of 13·5 m. (46 ft), accessible to vessels of 65,000 tons deadweight. Port equipment comprises a wharf, connected to the shore by a pier 400 m. (440 yds) long (Plate 1); the wharf includes a mineral berth 245 m. (265 yds) long and an oil berth 180 m. (195 yds) long. As a result of dredging operations carried out in 1966, ore-carriers up to 100,000 tons deadweight can now be accommodated. The operating system of the port ensures rapid ship turnround: every day two mineral trains deliver 30,000 tons of ore to Point Central; each wagon is emptied in 66 seconds (i.e. 4,000 tons of ore are unloaded per hour), and the ore is conveyed mechanically to a storage area with a capacity of $1\frac{1}{2}$ million tons. After screening – a process in which the sieve is adjusted according to the demands of the market – the ore is moved by conveyor belt to the quayside where loading gear allows both large and small vessels to be supplied whatever the state of the tide. The loading equipment can handle 3,000 tons an hour, a figure likely to be doubled in the near future. Vessels are thus dealt with very rapidly: 228 ore-carriers were loaded in 1964, 265 in 1965, 292 in 1966, 283 in 1967, and 271 in 1968. A total of 294 ships (ore-carriers and oil-tankers) entered Point Central in 1967, with a total net registered tonnage of 2,726,800 tons. The record loading figure of 54,000 tons achieved in December 1964 by the *Vronti* was surpassed in March 1968 when the *Sig Silver* took on 93,076 tons of ore bound for Japan.

Between 1963 and 1968 iron-ore exports through Nouadhibou rose from 1,300,000 tons to 7,703,152 tons (Table 3.3), thus reaching in four years the level of 7,500,000 tons at which MIFERMA intends to stabilise its annual output in the foreseeable future. This rapid growth has been stimulated in part by the considerable recent expansion of the iron and steel industry in industrialised countries, and in particular by the creation of iron and steel complexes located in port areas supplied with high-grade ores from relatively newly-developed sources in countries such as

Mauritania, Liberia, Brazil, Chile and Venezuela. The rapid development of iron-ore extraction and exportation in Mauritania has been made possible by a combination of circumstances: the high grade of the ore, the

Table 3.3

Exports of iron-ore from Point Central (Nouadhibou), 1963–8
(in tons)

Volume exported

1963 (6 months)	1,300,000
1964	4,980,000
1965	5,965,000
1966	7,157,300
1967	7,447,846
1968	7,703,152

Distribution by principal markets

	1968
Britain	1,910,739
France	1,472,991
Belgium	1,297,117
West Germany	1,292,084
Italy	974,935
Japan	540,688
Spain	188,298
U.S.A.	26,660

Source: *MIFERMA – Informations.*

facilities developed for bulk evacuation through a deep-water port, the relative proximity to major industrial areas in western Europe, and the careful regulation of the quality of the exported product in accordance with the specific demands of the consumers. The four countries with financial interests in MIFERMA – Britain, France, West Germany and

Italy – are themselves the chief consumers, but the proportion of the total output exported to these countries has decreased from 94 per cent in 1963 to 74 per cent in 1968. Meanwhile, tonnages despatched to Belgium, Japan, Spain and the U.S.A. have increased (see Table 3.3).

Nouadhibou is now therefore well established as an iron-ore exporting port, and is well equipped to become a major fishing port as well. The outstanding question now is that of the future of the port when the resources of high-grade iron ore are exhausted. Present indications are that exports of ore will continue, for detailed prospecting has revealed further reserves of magnetite in the Zouérate area. Efforts are also being made to improve the quality and profitability of the relatively low-grade iron ores of the Kédia d'Idjil area. Certain optimists see Nouadhibou as a future regional focus for the whole of northern Mauritania. It is hardly necessary to recall, however, that the vast Saharan area within the potential hinterland of the port is empty in the extreme; less than 10 per cent of the population of Mauritania is located there, and life is frugal. The least austere region within the potential tributary area of the new port – the Adrar region, where several oases are concentrated – is more easily linked with Nouakchott by an all-weather track than with Nouadhibou via Atar and Choum. The export of copper ore mined at Akjoujt (Fig. 3.1) began in 1969 through Nouakchott, although originally the use of Nouadhibou had been envisaged. Although the Mauritanian capital is situated on a low, inhospitable coast, it has been equipped since June 1966 with a wharf capable of handling 50,000 tons of cargo a year. In 1968 this figure was raised to 125,000 tons to accommodate exports of copper concentrates handled by the Société Minière de Mauritanie (SOMIMA). General cargo traffic handled at Nouakchott is extremely small – in 1967 exports totalled 609 tons and imports 33,888 tons (chiefly cement, rice, sugar and tea).

The future of Nouadhibou appears to rest, therefore, upon a fuller exploitation of its two basic resources, fish and iron ore. It has been indicated above that the development of an integrated fishing industry necessitates the establishment of a coherent policy. In the context of iron ore, the development of an integrated iron and steel mill at Nouadhibou is projected by the Mauritanian Government – one of several such projects on the West African coast. Coke would be brought from Europe by ore-carriers as return freight, limestone flux would be obtained from the enormous deposits of seashells on local beaches, and water supplies would be obtained either from the Bou Lanouar borehole or by means of a sea-water distillation plant. Such a scheme would represent a

pinnacle of achievement in the industrial and port expansion of
Nouadhibou, but existing conditions in terms of the location of the port
in relation to prospective market areas do not encourage an over-optimistic
view of future trends.[1]

1 *Further information on Nouadhibou is to be found in R. J. Harrison Church, 'Port-*
Étienne, a Mauritanian Pioneer Town', Geographical Journal, vol. 128 (1962)
498–504; T. Monod, 'Port-Étienne', Revue Maritime (1924) 442–72; R. J.
Peterec, 'Port-Étienne, le nouveau port international de la Mauritanie', Cahiers
d'Outre-Mer, vol. 16 (1963) 303–12; Société Industrielle de la Grande Pêche,
Mémoire sur la création d'une station de pêche à Port-Étienne (*Paris, 1931*);
C. Toupet, 'Les grands traits de la République Islamique de Mauritanie', L'infor-
mation géographique, vol. 26 (1962) 47–56; and 'Improvements at Port-Étienne:
New Jetty and Ore-handling Plant', Dock and Harbour Authority, vol. 44,
no. 519 (1964) 272–7. Technical details are provided by the following periodicals:
MIFERMA-Informations (*Paris, twice yearly*), Industries et Travaux d'Outre-
Mer (*Paris, monthly*), *and* La pêche maritime (*Paris, monthly*).

4. The Changing Role of the Port of Dakar[1]

Assane Seck

M. Assane Seck, who holds the Diplôme d'Études Supérieures in geography, was formerly on the staff of the University of Dakar and is now Minister of Education and Culture in the Senegalese Government. He is joint author (with Alfred Mondjannagni) of L'Afrique Occidentale (Paris, 1967).

The port of Dakar stands today as one of the most important seaports of tropical Africa. Some 5,455 ships, registering over 35 million tons, entered the port in 1967, and cargo movements exceeded 5 million tons. The hinterland of the port, although currently undergoing some contraction, remains larger than that of any other West African port. Dakar owes its importance to a variety of factors: to its situation at the most westerly point of the African continent, in relation to Atlantic trade routes; to the excellence of its sheltered, deep-water site, selected and developed by the French for strategic as well as economic motives; to its extensive and efficient rail and road links with its hinterland, which includes parts of Mauritania and Mali as well as Senegal itself; and finally to the wide range and high level of equipment within the port itself.

Dakar is in fact one of the best-equipped ports in the whole of tropical Africa. Approaches are excellent, with natural depths ranging from 10 to 12 m. (33 to 40 ft), and five lighthouses with ranges varying between 12 and 45 km. (8 and 31 miles). The harbour, enclosed within two large breakwaters, has an area of 216 hectares (530 acres), and adequate but irregular depths (see Fig. 4.1) which are maintained with little difficulty since sand movements are slight and the sea-water is relatively free from mud. This represents an unusual advantage for Dakar, and one shared by few ports elsewhere, since vessels can depend with certainty upon the depths shown on marine charts. In addition, the pattern of depths is such that vessels can turn easily inside the harbour, within a 65-hectare (158-acre) zone dredged to 10 or 11 m. (33–36 ft) and there is generally no need to follow specific channels when approaching berths.

1 *This chapter is based in part upon the author's thesis for the Diplôme d'Études Supérieures, in course of publication under the title* Dakar: métropole ouest-africaine. *The reader is referred also to a recent study by R. J. Peterec,* Dakar and West African Economic Development (New York, 1967).

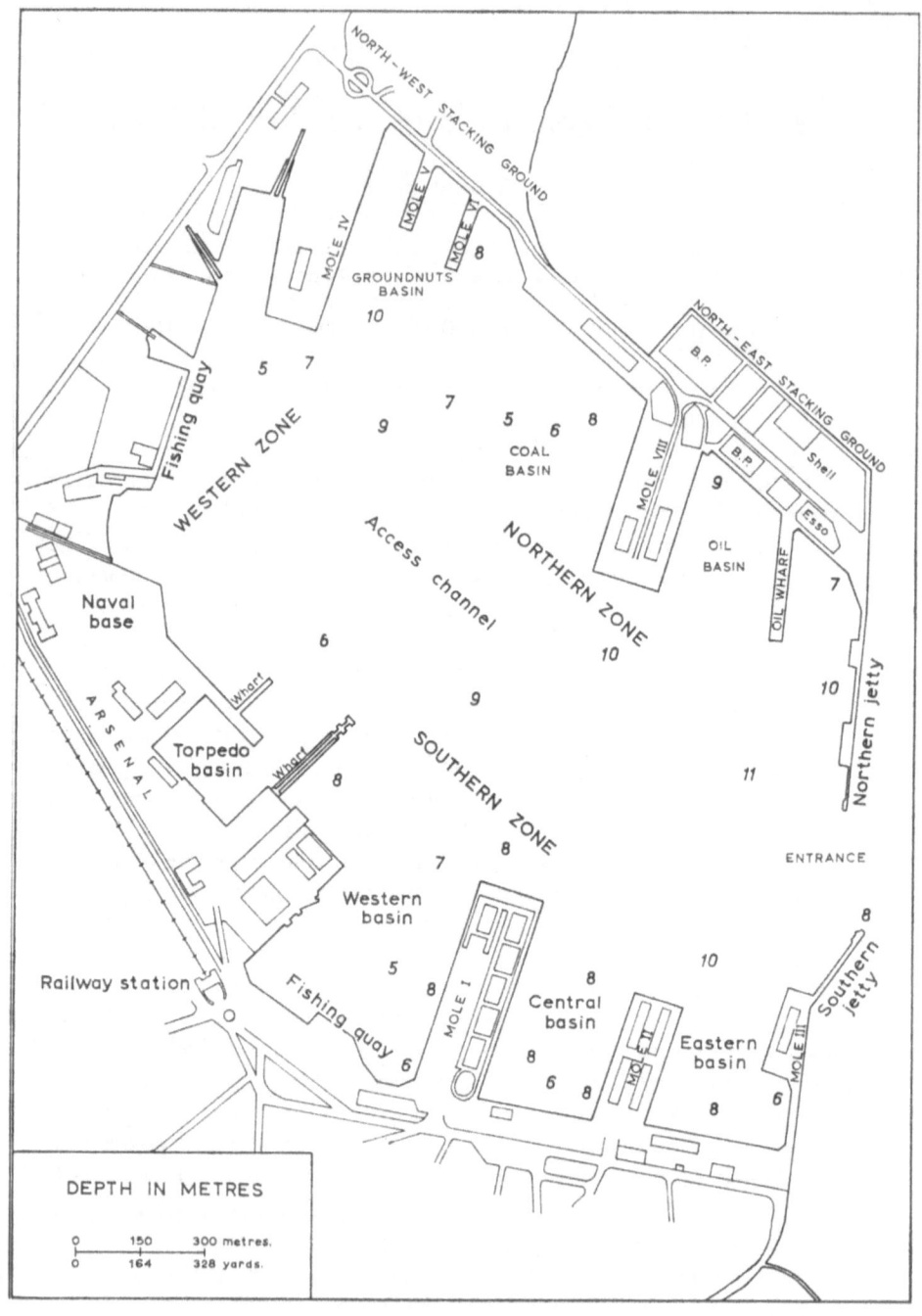

Fig. 4.1   Layout and facilities of the port of Dakar

Cargo-handling facilities in the port are well laid out and quite adequate to cope with current traffic flows. Total quayage amounts to 7,848 m. (8,554 yds), of which 3,008 m. (3,279 yds) has a depth alongside of 10 m. (33 ft), and only 1,263 m. (1,380 yds) has a depth of less than 5 m. (16 ft). Some forty-six berths are available, of which many are equipped with fuel-oil pumps and all have at least two water hydrants. Behind the quays the port is equipped with some 204,000 sq. m. (2,200,000 sq. ft) of stacking grounds and a total of 57,000 sq. m. (610,000 sq. ft) of covered transit sheds. There are also two cold stores, a dry dock belonging to the French Navy but also used commercially, and ship-repair facilities which allow Dakar to meet all modern shipping demands. Moreover, to these facilities may be added the not inconsiderable degree of efficiency maintained by the numerous companies providing port services both in physical terms within the port itself and in administrative terms from their headquarters in the town.

Port traffic at Dakar is organised on the basis of a three-zone system (Fig. 4.1) designed to avoid congestion. The first zone comprises the northern breakwater and its associated moles and transit areas; the north-western part of this zone handles phosphates, groundnuts, cereals, etc., and the south-eastern area deals with coal and oil cargoes. The second, western, zone lies between Mole IV and Mole I and is occupied mainly by repair workshops and the naval arsenal; commercial traffic within this zone is slight, but the fishing activity focused upon the western basin with its cold store is considerable. The third, southern, zone (Plate 2) is the oldest part of the port, the most active in terms of general cargo movements, and is located close to the commercial centre of the town; within this part of the port Mole III was designated a free zone for the Republic of Mali in 1963. The layout of the port thus reflects its three basic functions as a bunker port, a trade centre handling both specialised and general cargo, and a naval base. This paper examines the changing character of the first and second of these roles.

Dakar as a bunkering port

A high proportion of all vessels calling at Dakar do so for the sole or partial purpose of taking on supplies of fuel, food and water. The loading of fuel, frequently the most important of these operations, is the one which usually determines the location of bunkering operations within the port for a given vessel. Between 70 and 80 per cent of vessels entering the port take on provisions, and some 40 per cent enter solely for this purpose. If one includes the tankers bringing oil supplies to the port, the proportion of vessels directly associated with the bunkering function rises to over 90 per cent. In 1968, out of a total volume of cargo traffic of 5,871,534 tons, bunker cargoes comprised 2,028,944 tons (of which oil formed 1,433,746 tons and water 595,198 tons)[1] or 35 per cent of the total. The role of Dakar as a bunkering port is long-established, and was relatively more important in the past when local economic development was in its initial stages; in 1913, for example, bunker cargoes represented 83·3 per cent of the total traffic handled. The closing of the Suez Canal in 1967 proved to be greatly to Dakar's advantage; the tonnage of shipping entering the port in 1968 showed a 65 per cent increase compared with 1966.

Dakar is not, however, alone in developing a role as a 'service station' for this part of the Atlantic Ocean: Madeira, the Canary Islands and the Cape Verde Islands are strong rivals, but the degree of competition varies. Madeira is not well equipped and is too near to Europe to be an important competitor. The Cape Verde Islands have poor water supplies and modern port equipment is very limited. The strongest competition comes from the Canary Islands: situated less than sixty miles off the African coast, the archipelago lies directly on the important oil supply route between Central America and Europe. Costs of bunkering services there are in some measure offset by the cargoes of cheap local produce which are taken on. Both Grand Canary and Tenerife are equipped with modern deep-water ports, but inter-island competition is probably more important than competition from either of these islands with Dakar. The port of Las Palmas (on Grand Canary Island) is, however, an important rival. The bay which shelters the port is protected by a 2,600-m. (2,835-yd) breakwater (named after General Franco); depths alongside this break-water vary between 10 and 19 m. (33–62 ft) over a length of 2,000 m. (2,180 yds), and berthing conditions are generally better than at Dakar.[2] The largest oil-tankers can be received and, like Dakar, Las Palmas is an area of open competition between the various oil companies; but in technical terms Dakar is better equipped to service vessels efficiently with fuel oils. Nevertheless the benefit of a geographical situation at least as favourable as that of Dakar, together with the variety of additional export cargo and the tourist industry, tend to tip the balance in favour of Las Palmas, so much so that rivalry in the sphere of fuel-oil supplies is a constant source of anxiety to the Dakar authorities. This competition has

1 *Except where otherwise stated, all figures given in this chapter have been obtained from the Direction du Port de Dakar, notably from the annual publication* Statistique du Port de Commerce de Dakar.

2 Instructions nautiques, Série C (VI): Afrique – Côte Ouest, *vol. 1 (Paris: Service Hydrographique de la Marine).*

been an important influence upon the growth of Dakar, and specific physical developments have taken place in order to deal with it. Also, a dual tariff system exists which discriminates against those vessels participating in the commercial as well as the bunkering activities of the port; before independence (1960) commercial calls cost twice as much as bunkering calls in terms of pilotage and port dues.

There is, however, a sense in which the bunkering traffic is merely a prestige activity from which relatively little profit is derived. The immediate post-independence period has tended, moreover, to favour policies encouraging traffic links with the immediate hinterland of the port. Thus the tax systems introduced in January 1962 still favoured vessels calling exclusively for bunkering purposes in so far as pilotage costs were concerned, but harbour dues chiefly affecting vessels involved in commercial operations were not increased. In addition, whilst taxes on essential commodities such as coal, oil, rice and sugar imports and groundnut and phosphate exports have remained virtually unchanged since 1953, a discriminatory tax structure has been introduced for many products designed specifically to favour items of local origin. Thus whilst phosphates, vegetable oils, sweets and matches of local manufacture are charged 180 francs per ton on export, and local cement is charged 120 francs per ton, the same goods (when their import is permitted) are classified as 'other merchandise' in the import tariff structure and are taxed at 250 francs per ton. In the same way wheat flour from local mills pays only 20 francs a ton on export, whilst imported flour is charged at 110 francs per ton. Thus the port of Dakar is encouraged to become increasingly an instrument of local economic development as well as a port of significance in a much wider context.

Dakar as a commercial port

In addition to its role as a bunkering port, Dakar is an important focus for traffic in a wide variety of raw materials and manufactured goods. This function has allowed Dakar to develop a vast hinterland, but relationships between port and tributary area have varied with political circumstances.

Shipping movements

Vessels belonging to almost 200 companies call regularly at Dakar. Many of these vessels do not terminate their voyage at Dakar; in general, French vessels serve the entire West African coast, particularly those ports situated in former French territories, whilst British vessels similarly tend to call at former British ports. Scandinavian vessels, specialists in the tramp trade, are frequent visitors, whilst numerous vessels serving primarily the trade of areas within the hinterland of Dakar use the port as their terminus or point of departure.

The northern zone of the port (Fig. 4.1) is used by five groups of ships which visit the port regularly, carrying respectively oil, minerals, ground-nuts and their derivatives, cereals and wine. The oil-tankers come mainly from Central America and carry refined products, imports of which reach quite a high level at Dakar. Between 1951 and 1954 the total tonnage of oil products unloaded in the port exceeded 1 million tons annually, and rose sharply to 1·6 million and 2·1 million tons in 1956 and 1957 respec-tively owing to the Suez crisis. This pattern was repeated following the closure of the Suez Canal in 1967; bunkering rose from 810,000 tons in 1966 to 1,433,746 tons in 1968. Since the opening of the M'Bao refinery near Dakar in 1964 (capacity 1 million tons per annum) imports of crude oil have increased sharply, particularly from Algeria and Gabon, since the Senegalese authorities decided that as far as possible crude oil should be purchased within the franc zone. Thus, in the first half of 1964, the M'Bao refinery imported 190,000 tons of crude oil – 115,000 tons from Algeria and 75,000 tons from Gabon. In 1966 the refinery treated 448,128 tons, mainly from Algeria. All oil products consumed locally (334,995 tons in 1966) are provided by the M'Bao refinery.

Vessels transporting exports of phosphates from Senegal form a second group whose numbers are likely to increase with the expansion in the exploitation of the calcium phosphate deposits of the Taiba area, 40 km. (25 miles) north of Thiès (Fig. 4.2). These phosphates are shipped from Mole V, and aluminium phosphates from Thiès are loaded at the adjacent quayside. In 1968 phosphate exports totalled 1,109,893 tons, or 65 per cent of all cargo embarkations.

A third group is formed by ships carrying groundnuts and their derivatives. Until quite recently, groundnut products constituted by far the most important commodity group exported through Dakar. Today their place has been taken by mineral products, not only because mineral exports have risen but also because since the Second World War the groundnut-processing industry in Senegal has advanced so considerably that tonnages of unprocessed groundnuts for export have declined sharply. Groundnuts and their products now occupy third place, therefore, after oil products and phosphates, in Dakar's list of exports. These changes have also affected the pattern of shipping movements at Dakar. Before the Second World War the greater part of the groundnut harvest was exported unprocessed, and the need to ship the entire stock from the quayside stacking grounds during the dry season (November to May) meant that a large number of vessels used the port during that part of the year. These vessels included both large cargo vessels unable to reach the smaller river ports of Senegal and medium-sized tramps only partially loaded elsewhere. Today, with the development of shelling and crushing mills in Senegal itself, the number of ships carrying unprocessed ground-

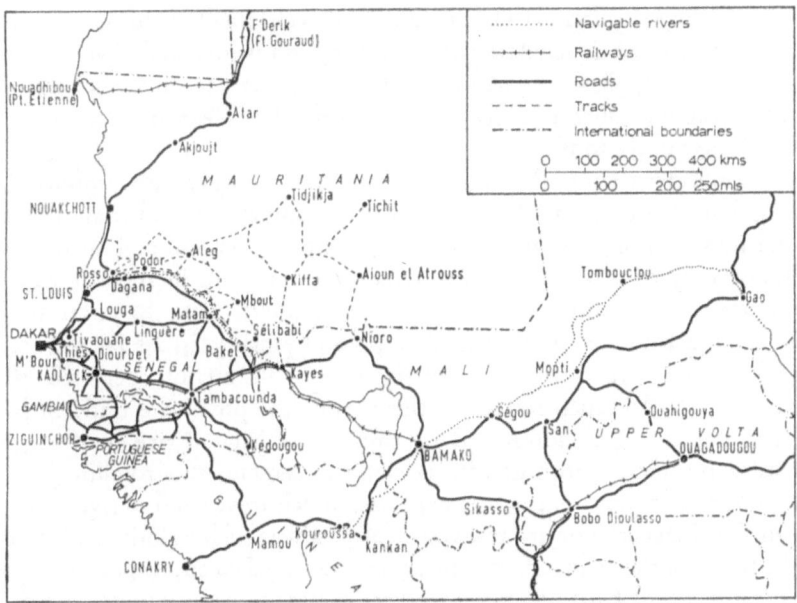

Fig. 4.2   The hinterland of the port of Dakar

nuts has decreased greatly, although the groundnut harvest continues to rise. On the other hand, vessels of a different type now take on cargoes of groundnut oil and oil cake, and the traffic has become much less seasonal than previously.

The importing of cereals to supply the flour mills of Dakar gives rise to an important fourth group of vessels using the port. Wheat and maize occupy an important place in the Senegalese economy, for not only do they stimulate traffic movements but they also provide the basis for an important industry which in turn gives rise to a not inconsiderable flow of exports. Until 1962, imports of cereals were gradually increasing – from 80,182 tons in 1957 to 150,136 tons in 1962. The greater part of the production of the Dakar flour mills was re-exported in various forms, and the tonnages despatched in this way rose to 76,551 tons in 1962. The recent development of a flour-milling industry at Abidjan has, however, diminished both imports and exports of cereals and their products at Dakar: in 1965 cereal imports stood at 121,408 tons and re-exports at 22,434 tons.

Tankers carrying wine form the last, and least important, group of vessels calling regularly at Dakar. Only relatively small quantities of wine are imported, for Senegalese wine consumption is low since over 75 per cent of the population is Moslem. Wine imports have never exceeded 20,000 tons and since 1962 have only marginally exceeded 10,000 tons a year (1965: 9,233 tons).

The northern zone of the port is thus economically very important, being linked closely with local agricultural, mineral and industrial output; total traffic within this zone (excluding oil products, most of which are immediately re-exported) rose to over 1,600,000 tons in 1966 of which imports formed only 150,000 tons. The southern zone, in contrast, is above all a zone handling general cargo and passengers. As far as cargo traffic is concerned there is an important distinction between the Senegalese and Malian sectors of this part of the port. The trade of the Senegalese sector is dominated by imports of bagged rice, loaf sugar, textiles, machinery, iron goods, sheet metal, cement, etc., totalling some 600,000 tons a year; export tonnages are much lower and derive chiefly from local factories or are re-exports of products such as hides and skins. The traffic of the Malian free zone is likewise unbalanced: imports include a great variety of manufactured and semi-manufactured goods (tractors, machinery, sheet metal, etc.), and exports include some important raw materials such as groundnuts, cotton, gum, hides, etc. In 1964 import traffic in the free zone[1] rose to 62,728 tons and exports to 28,011 tons (of which two-thirds were groundnuts); in 1965 imports rose to 80,072 tons, largely as a result of increases in cotton-textiles and sugar, but exports showed a slight tendency to decline (24,626 tons), owing mainly to increasing local processing of groundnuts.

Passenger traffic through the port remains considerable, in spite of competition from air transport. In recent years total passenger movements (arrivals and departures, including passengers in transit) have stood around

---

1 *The Malian free zone began operating in 1964. Not all Malian imports from Senegal pass through this zone, since certain goods originate as withdrawals from Senegalese stockpiles or from the industrial production of Senegal itself (e.g. cement, petroleum products).*

the 100,000 mark, although both departures and numbers of passengers in transit show a general tendency to decline. Both passenger and commercial traffic handled in the southern zone of the port are largely effected by vessels serving the entire West African coast, but two other types of vessel are characteristically linked with this section of the port: liners serving the Dakar–Marseilles route, and small coastal vessels linking Dakar with other West African seaports. Three liners belonging to the Compagnie Paquet – the *Lyautey*, the *Djenné* and the *Ancerville* – link Dakar with Marseilles; traffic on this route is heaviest in the northerly direction at the beginning of the rainy season (June), when many Europeans go on vacation, and in the reverse direction the ships are always fully booked in October and November just as the main period of Senegalese economic and administrative activity begins. Coastal vessels serving the smaller ports of Senegal and other West African countries include the *Ouolof* (1,709 tons) belonging to the Compagnie Paquet, and vessels owned by the Société Navale d'Importation et d'Exportation (S.N.I.E.) – the *Diorhane* (2,100 tons) and the *Avondiré* (1,750 tons). These vessels load cargoes of salt at Kaolack (Fig. 4.2) and, after taking on supplies and perhaps unloading part of their cargo at Dakar, take on additional cargo and call at various ports along the Gulf of Guinea (Sassandra, Abidjan, Lomé, Cotonou and sometimes Douala). Their return freight usually consists of timber, chiefly destined for Dakar. A third vessel belonging to the S.N.I.E. company, the *Saint-Honorat* (300 tons), links Dakar with the small ports along the Senegal river.[1] These various vessels are all based at Dakar and are manned by African crews with European senior officers.

The maritime links maintained by these various types of vessel formerly followed a simple twofold system of operation within the trading economy of the area, with a period of considerable commercial and transport activity during and immediately after the harvest season and a period of relative inactivity during the rains. This pattern has by no means completely disappeared and is still reflected in the seasonal variations in the numbers of vessels entering Dakar for commercial purposes, variations which are themselves reflected in higher-level imports of petroleum products during the more active season. On the other hand, excluding bunkering supplies, imports show a very clear peak at the beginning of

1 Études d'estuaires au Sénégal – Partie Saloum, *Bureau Central pour les Équipements d'Outre-Mer*, Dec. 1960. *Until 1960 the S.N.I.E. company operated a fourth vessel, the* Soulac, *which on account of its age (fifteen years) and heavy repair costs was transferred to the Messageries du Sénégal company at St Louis and now serves on the Senegal river.*

C

the year which is closely linked with the most active season in the commercial and financial spheres. But the seasonal nature of the traffic flow is much less strongly marked today in terms of exports as a result of the introduction of stabilising factors such as the more even distribution throughout the year of groundnuts and their derivatives, the export of flour from the Dakar mills, and above all the movement through the port of large quantities of locally produced phosphates.

Patterns of export commodity flow through Dakar

Groundnuts provide the key to an understanding of patterns of export commodity flow to and through the port of Dakar; movement of other commodities generally follows a similar pattern, if only because groundnuts, as the dominant export crop throughout the hinterland area, are intimately linked with the pattern of surface communication, and not all non-groundnut-producing areas have adequate road and rail networks. Two main types of commodity are involved in the flow towards Dakar from the hinterlands, agricultural products and minerals. All are destined either for export (in crude form or after partial or complete processing), or for processing for local consumption in the Dakar area and throughout the hinterland.

A wide variety of agricultural exports passes through the port – groundnuts and their derivatives, cotton, gum arabic, hides and skins, raw wool, kapok, etc. – but groundnuts are still overwhelmingly predominant and account for over 95 per cent of the tonnage of all agricultural exports. In 1966 Dakar exported 415,800 tons of groundnut products, comprising 167,400 tons of decorticated groundnuts, 142,200 tons of oil cake, 100,200 tons of groundnut oil, 4,000 tons of groundnut bran and 2,000 tons of groundnut flour. Taken together these products represent some 500,000 tons of undecorticated groundnuts, or approximately 25 per cent of the total commercial output of the entire groundnut zone which extends from Senegal to northern Nigeria south of the 500-mm. (20-in.) isohyet. Within Senegal, not all the groundnuts grown for export pass through Dakar – Kaolack and Ziguinchor export directly at least a part of their local output – but the primary port remains the predominant means of evacuation. In 1964, 1965 and 1966 the equivalent of undecorticated groundnuts exported through Dakar totalled respectively 500,000 tons, 456,000 tons and 540,000 tons; combined flow through the other Senegalese ports was respectively 253,000 tons, 320,100 tons and 310,000 tons.

Only a relatively small part of the export production from the non-Senegalese areas of the West African groundnut zone passes through

Dakar, for reasons associated primarily with the economics of transportation but also with the political situation. The nearest foreign source of groundnut exports is Gambia, whence the production is despatched via the Gambia river and the port of Bathurst; the peculiarities of the political geography of Senegambia largely rule out inter-port competition in this area. Similarly, since independence, groundnut production from the frontier areas of northern Guinea has been channelled through Conakry rather than Dakar. Production from Upper Volta, the Niger Republic and northern Nigeria can, of course, be more conveniently evacuated through ports on the south coast of West Africa. Amongst the inland states of West Africa only Mali exports groundnuts through Dakar. The movement of Malian groundnuts through Senegal has had a complex evolution, for they were channelled originally through St Louis, and later through Dakar and Kaolack; Dakar eventually acquired a monopoly which was broken by the collapse of the Senegal–Mali Federation, which had the effect of diverting all Malian traffic away from Dakar and towards Abidjan between 1960 and 1963. The eventual solution of this political crisis has meant that at least part of the Malian groundnut crop is once more exported through Dakar. In the 1964–5 season, Mali exported a total of 30,000 tons of groundnuts, 18,000 tons through Dakar and 12,000 tons through Abidjan.

Other agricultural products exported through Dakar do not normally total more than about 10,000 tons a year. Some of these minor items originate in groundnut-producing areas and use the transport media primarily developed for the predominant crop; but more distant regions and more valuable products (per unit of weight) are also involved. Important items include cotton, shea-nut butter and raw wool – all of which come exclusively from Mali – and also hides, skins and gum arabic which originate in northern Senegal, Mali and Mauritania. The relative importance of these various items in relation to groundnuts can be seen in the following breakdown of agricultural exports through Dakar in 1966: groundnut products, 415,800 tons; gum arabic, 4,624 tons; raw cotton, 2,740 tons; hides and skins, 1,718 tons; raw wool, 220 tons; kapok, 20 tons. The total volume of agricultural exports (427,122 tons in 1966) is very unevenly distributed between Senegal (407,935 tons), Mali (17,187 tons) and Mauritania (2,000 tons), and is moreover greatly exceeded by the volume of mineral exports.

Mineral exports through the port of Dakar consist essentially of Senegalese products – the aluminium phosphates from Lam-Lam, 12 km. ($7\frac{1}{2}$ miles) from Thiès, and the calcium phosphates from Taiba, near

Tivaouane. Exports of phosphates through Dakar have increased very rapidly since 1959, when less than 100,000 tons passed through the port; by 1963 this figure had risen to 537,216 tons, and in 1966 the total reached 1,156,468 tons. This rapid growth is due primarily to the expansion of output from Taiba.

In general terms, therefore, the port of Dakar draws its export commodities primarily from the groundnut-producing areas of Senegal (except the Casamance region south of Gambia), from northern Senegal and southern Mauritania, which provide gum arabic and most of the hides and skins exported, and from the western half of Mali, which contributes groundnuts, cotton and miscellaneous minor items. It is through the port of Dakar that these areas maintain their export trading links with non-African countries, but the services rendered by the port in terms of the distribution of imported goods are even more important. Many manufactured products imported at Dakar are better able to withstand considerable transport costs than the primary products which originate within the hinterland.

Dakar is virtually the only primary distribution centre for manufactured goods in Senegal, partly as a result of the progressive elimination of St Louis and Rufisque as overseas trade centres. Changes in the traditional pattern of traffic flow through Senegal have also occurred as a result of the disintegration of the Senegal–Mali Federation, the subsequent reduction of Malian traffic at Dakar, and the associated creation in 1960 of the Office de Commercialisation Agricole (O.C.A.). As a result of these factors the two regional ports of Kaolack and Ziguinchor, which still preserve in a limited way their role as exporting ports, play an insignificant part in the import trade. Most of the cargoes which in earlier years would have been unloaded at the southern ports now pass through Dakar, which has accordingly become virtually the only importing port in Senegal. In 1968 a total volume of 2,136,024 tons of goods (1,207,611 tons of petroleum products and 928,413 tons of general cargo) were unloaded at Dakar; these figures are now higher than those achieved during the immediate pre-independence period, before the political structure of former French West Africa was changed (in 1958 total imports were 1,770,595 tons, including 1,070,885 tons of petroleum products and 699,710 tons of general cargo).

Certain legal and administrative changes have been necessary, as a result of the emergence to separate independent statehood of the various units of former French West Africa, in order to protect the developing industries of the Cape Verde peninsula (Fig. 4.3). The guiding principle

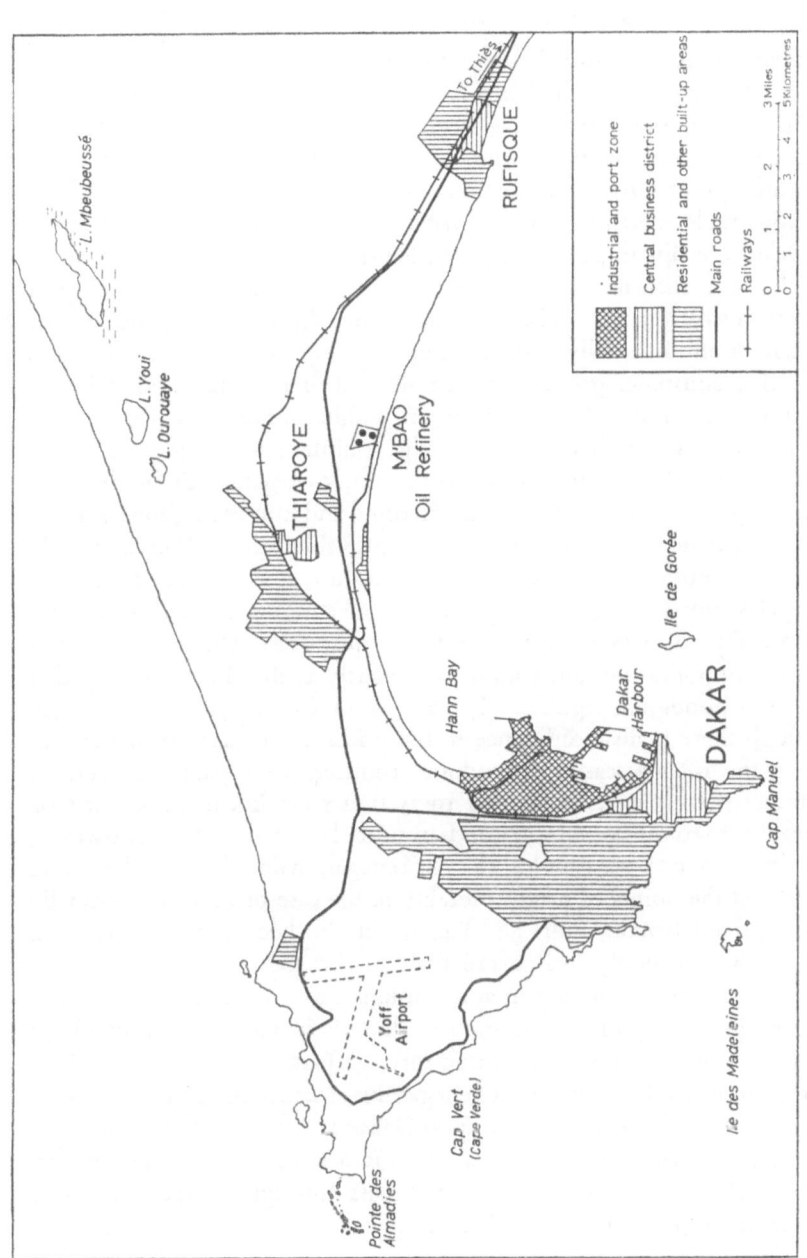

Fig. 4·3   The site of Dakar

involved in these measures has been the prohibition of imports of all goods the local demand for which can be satisfied by Senegalese industries; in addition, a quota system has been introduced under which limited authorised consignments of goods may be admitted where local production is insufficient or where competition between local and imported products is considerable.[1] As a result of this protective system, the industrial products of Dakar are distributed throughout Senegal, not only within the areas which export their produce through Dakar but also throughout the export hinterlands of Kaolack and Ziguinchor. Thus the whole of Senegal now finds itself almost totally dependent upon Dakar as a source of imported and other consumer goods. Goods unloaded or manufactured at Dakar are, however, destined for distribution within an area which transcends not only Senegal itself but also the export hinterland of the port.

The role of Dakar in relation to its extra-Senegalese hinterland, as a primary distribution point for manufactured and imported goods, is more important than its role as a collecting centre for exports. Gambia clearly forms an anomaly, for Gambian exports are not evacuated through Senegalese ports and goods originating in or entering through Dakar are not distributed in Gambia in significant quantities. This is not only a question of political boundaries; it is primarily a reflection of the fact that imported goods are significantly cheaper in Gambia than in Senegal, although there is little difference in the prices of primary exports. Thus during the 1966–7 season groundnut producers were paid, on average, 19 francs per kg. in Gambia and 19·75 francs per kg. in those parts of Senegal near to the boundary; goods imported through Gambia, however, cost on average 33 per cent less than in Senegal, mainly because they were cheaper at the points of origin (notably in the case of imports, especially textiles, from Britain, India and Japan) but also because import taxes in Gambia are generally less severe than in Senegal. As a result of this situation, goods originating in and through Dakar cannot penetrate into Gambia in significant quantities, and there is in fact a vigorous illegal traffic in Senegal in goods imported through Gambia.

In contrast to the situation in Senegambia, Mauritania and Mali are to a considerable extent dependent upon Dakar as a source of manufactured goods. In the case of Mauritania, as shown elsewhere in this volume, exports of iron ore and fish do not pass through Dakar but utilise Nouadhibou (Port Étienne); Dakar remains, however, the virtually

1 Competition of this kind is not always a serious problem, and no protection is necessary for certain local products such as groundnut oil and cement, which are widely available and the prices of which are lower than those of the corresponding imported items.

uncontested point of entry for manufactured goods. Until the construction of the railway linking the iron-ore deposits of northern Mauritania with the coast, Nouadhibou was relatively isolated from the rest of Mauritania and, in the absence of direct regular links with non-African industrialised countries, received most of its supplies via Dakar. The remainder of Mauritania was even more closely dependent upon Dakar (except for the extreme northern area which traded to some extent with Morocco), to which it was linked by road, rail, river and sea routes in a variety of combinations. The recent development of the iron-ore resources of the F'Derik (Fort Gouraud) area (see Chapter 3) has modified the situation somewhat as far as Dakar is concerned. All equipment and supplies for the iron-ore development project have been obtained from Europe and have utilised Nouadhibou in preference to Dakar as a point of entry; the mineral railway now permits Nouadhibou, rather than Dakar, to supply the greater part of northern Mauritania. However, a considerable part of Mauritania is still supplied by overland transport from Dakar, and the flow of goods from Dakar through Nouadhibou itself still continues; petroleum products, rice, flour and a variety of other items reach Mauritania by this route. Plans to extend the recently constructed wharf at Nouakchott could well divert some traffic through that port also.

The Republic of Mali also receives the greater part of its imports through the port of Dakar; the most common commodities involved are cement, food and drink, tobacco and cigarettes, textiles and a variety of equipment. In addition to imported goods entering through the port of Dakar, Mali also imports many of the products of Dakar's industrial zone; flour, sugar, cement and petroleum products in particular. Trade between Senegal and Mali in these items was formerly very vigorous but has recently declined, partly as a result of competition from Abidjan and partly because of the obstacle provided by the non-convertible Malian franc. Cement imports into Mali from Senegal exceeded 25,000 tons in 1960, and in 1963–4 stood at 18,793 and 22,171 tons respectively as a result of payments in kind made by Senegal when the Senegal–Mali Federation broke up; by 1965, however, they had fallen to 9,435 tons, whilst more than 30,000 tons of cement reached Mali by other routes. The same problem has arisen in the case of flour imports into Mali from Senegal, which declined from 22,000 tons in 1958 to 12,000 tons in 1963. In contrast, Dakar is still the chief source of petroleum products for the Malian market, formerly supplied by means of direct importation and now provided through the M'Bao refinery; a total of 39,000 tons was despatched to Mali in 1965, compared with 30,000 tons in 1958. Altogether, Mali received from Dakar (from the industrial area and via the port) a total of 130,089 tons of merchandise in 1966 (132,072 tons in 1965), valued at 2,789 million Malian francs (3,748 million in 1965). Malian imports via Abidjan totalled only 40,351 tons in 1966 (43,036 tons in 1965), worth 662 million Malian francs (1,028 million in 1965).

Dakar thus provides an excellent example of a tropical African colonial seaport currently responding to new political and economic forces. In the post-1960 period the traditional role of the port of Dakar, and the colonially induced transport patterns of former French West Africa, have undergone considerable modification. Malian dependence upon Dakar has been diluted, as links with Abidjan have strengthened; and the historic relationship with Mauritania is in process of partial severance as Nouadhibou and Nouakchott develop. Although the areal extent of Dakar's hinterland is thus being reduced and the relative importance of the port is declining, the absolute volume and value of cargo movements are likely to continue to increase as the Senegalese economy expands and as Dakar continues its role as one of the world's great bunkering ports. The import hinterland of the port, more extensive than the export hinterland, continues to include the whole of Senegal together with much of Mauritania and Mali. Dakar is still a great port, despite the contraction of its formerly immense hinterland, and is still playing a vital role in the economic development of the northern zone of West Africa. Whilst these developments have occasioned a certain atmosphere of crisis in recent years, the changes that are being experienced are part of a progressive evolution rather than a regressive devolution, and Dakar in the future is likely to emerge as a true economic growth pole and not simply as a maritime service station.

5. Physical Potential and Economic Reality: the Underdevelopment of the Port of Freetown

J. McKay

Mr John McKay, M.A., spent the academic year 1964–5 at Fourah Bay College working on the commercial structure of Freetown. He was Assistant Lecturer in Geography at Liverpool University from 1965 to 1967, when he was appointed to the staff of the Bureau of Resource Assessment and Land Use Planning in the University College of Dar es Salaam.

The considerable physical advantages that Freetown possesses as a port site have long been recognised, as the accounts by numerous early seafarers bear witness:

> This noble river is at least two leagues wide at its entrance, and has a safe and deep channel for ships of any burthen, and affords excellent anchorage at all seasons. It continues the same breadth for six or seven miles . . . On the north side of Sierra-Leone river the land is low and level . . . but on the south side it rises into hills, which, forming one upon the other, tower into lofty mountains crowned with perpetual verdure. From the foot of these hills points of land project into the sea, which form excellent bays for shipping and craft, and convenient places for hauling the seine.[1]

The estuary of the Sierra Leone river stands out above any other natural port site on the whole of the West African coast, generally made inhospitable to shipping by its heavy surf and shallow, muddy creeks. The area of sheltered water of Freetown's estuary is so large that on one famous occasion during the Second World War, when the harbour was an important convoy staging post, no fewer than 250 ships were simultaneously anchored there. The town and the port have grown up on the southern shore of the river, where a series of raised beaches cutting into the remnant lopolith of the Peninsula Mountains[2] provides good building land. From the point of view of the port, this site is made doubly attractive by the strong current which sweeps this shore, ensuring that the deepest part of the harbour is close to the land. At the same time the tidal range is

---

1 J. Mathews, A Voyage to the Sierra Leone River (London, 1788) pp. 20–1; reprinted 1968.
2 S. Gregory, 'The Raised Beaches of the Peninsula Area of Sierra Leone', Trans. Inst. Br. Geogr. vol. 31 (1962) 15–22.

relatively low [spring 3–3·4 m. (10–11 ft), neap 2·1–2·7 m. (7–9 ft)] and does not usually inhibit the use of the port.[1] Access to the estuary is by a 1·2-km. (¾-mile)-wide harbour entrance, where the depth of water is about 11 m. (36 ft).

Early explorers of the West African coast were often grateful for the sheltered anchorage that Freetown's harbour provided, but of equal importance to them was the reliable source of fresh water found at a point near the centre of the present town. Hence 'King Jimmy's Watering Place', named after a local ruler, became a favourite provisioning point. However, this particular water supply no longer exists. A physical characteristic of greater contemporary importance is the swamp-free path to the interior of the country that is formed by the northern edge of the Peninsula Mountains. This corridor is utilised by the road and railway which are Freetown's main links with its hinterland.

The raised beach area already mentioned constituted a flat yet well-drained site for the town, which was founded at the end of the eighteenth century as a home for former slaves from the New World. In 1787 the first group of settlers founded 'Granville Town', named in honour of their benefactor Granville Sharp.[2] This community endured tremendous hardships, and the new town did not become firmly established until other groups arrived from Nova Scotia and Jamaica. After 1808, when an Admiralty Court was established there, the population was swelled by freed slaves taken from slaving ships captured along the Guinea coast. From this time the population increased steadily to the present figure of 127,917 (1963 census). The town has expanded from its original centre at Fort Thornton (now State House) to extend over 6½–8 km. (4 to 5 miles) of the raised beach.[3]

After reviewing all of these important site advantages which Freetown possesses, Jarrett concluded:

> *Indeed when we sum up in our minds the entire physical advantages which the site enjoys for both town and port, we cease to marvel, as the popular custom is, at the location of Freetown and wonder instead at the fine array of natural advantages assembled in this place . . .*[4]

How, he asks, could a port with all these physical attributes fail to prosper? Although, as will be seen, some port development has taken place, the

1 *J. I. Clarke, 'Ports', in J. I. Clarke (ed.), Sierra Leone in Maps (London, 1966) pp. 108–9.*
2 *C. Fyfe, History of Sierra Leone (London, 1962).*
3 *J. McKay, 'Freetown', in Sierra Leone in Maps, pp. 58–9.*
4 *H. R. Jarrett, 'The Port and Town of Freetown', Geography, vol. 40 (1955) 108–18.*

dominant impression that one has is of a superb natural site that is very much under-utilised. Far more activity is found on other parts of the West African coast, where large sums have had to be spent on providing artificial harbours. The aim of this essay is to analyse possible reasons for the comparative lack of port development at Freetown, and examine whether this represents a serious and avoidable waste of natural potentialities in that part of West Africa.

Inter-port competition in Sierra Leone

Taaffe, Morrill and Gould, as part of their analysis of the development of transportation networks within a number of developing countries, have pointed to the establishment, in the early phases of colonisation, of many small, closely spaced ports.[1] The Sierra Leone coast certainly passed through such a phase. During the nineteenth century European and Creole traders established trading factories at the heads of navigation of a number of the small creeks to the south of Freetown. Bonthe became the base of this trade, but for some time these factories played an important role in the export of palm products in particular.[2] In most African countries, as further development took place, one or two of these ports have outgrown the other competitors, have been able to establish stronger trading links with the interior, and thus have been able to monopolise the pattern of commerce. Freetown, partly because of its considerable natural advantages, has certainly far outgrown these early rivals.

Bonthe enjoyed a period of prosperity during the latter part of the nineteenth century. In 1871 the town and nearby York Island had a total population of 4,333, and its export trade was valued at about £100,000, but since then its population has increased but little and its trade steadily dwindled. In 1963 the town had a population of 6,230 and York Island 451. The port now ships less than $\frac{1}{2}$ per cent of Sierra Leone's export trade, mainly piassava and coffee, and its imports are negligible.[3] Part of the problem has been the silting of the old harbour in the Sherbro river, so that cargoes must now be lightered some 11 km. (7 miles) to the waiting ships. However, a major factor has been the loss of the export trade of the palm products originating in central Sierra Leone. The construction of the railway from Freetown, which reached Pendembu in 1908, siphoned

1 E. J. Taaffe, R. L. Morrill and P. R. Gould, 'Transport Expansion in under-developed Countries: A Comparative analysis', Geographical Review, vol. 53 (1963) 503–29.
2 A. M. Howard, 'Economic History', in Sierra Leone in Maps, pp. 74–5.
3 M. E. E. Harvey, 'Bonthe: A Geographical Study of a Moribund Port and its Environs', Bull. Jour. Sierra Leone Geog. Ass., vol. 10 (1966) 60–75.

off much of this trade. At the same time Bo became the provincial head-quarters of the Produce Marketing Board, and this attracted products to the railway and thence to Freetown.

The construction of this Sierra Leone Government Railway has allowed Freetown effectively to dominate the export trade of agricultural products from all parts of the country, and especially the coffee, palm-oil and cocoa produced in the Eastern Province. The railway was of a type familiar in many former colonial countries in Africa. It was built partly to allow easier exploitation of the interior areas and the export of primary products, and was also seen as having military value in that troops could be sent to any part of the northern and eastern frontiers where French expansion was feared. The military need was responsible for a hasty expansion of the network, but this urgency, along with the need to keep construction costs to a minimum, resulted in the choice of a very narrow gauge of 0·75 m. (2 ft 6 in.) and an alignment which in parts is extremely tortuous and steep. As a result the capacity of the line is low, and no train can exceed 32 km. per hour (20 m.p.h.).

In spite of these difficulties, the export trade of agricultural products developed, and this in turn was reflected in the physical development of the port of Freetown.

Table 5.1

Freetown: exports of major agricultural crops, 1900–45
(£'000)

	1900	1910	1920	1930	1935	1937	1939–45 average
Piassava	—	—	—	32	36	44	56
Ginger	—	33	60	57	37	98	53
Palm-oil	—	63	123	79	36	42	—
Palm kernels	172	645	1,400	450	584	885	381
Kola	—	192	627	186	39	61	35

Source: *Howard, op. cit. p. 74.*

For many years this ever-increasing volume of trade was transferred to ocean ships by lighters, and this resulted in the construction of a number of small jetties and wharves. Until 1954 the main port area was concentrated around the lighter quays of Government Wharf, which had been completed in 1875 on the site of a port area that had been used since the foundation of the town. Complaints had been made about the inadequacy of these facilities as early as 1901, when a report was presented on the need for a deep-water quay.[1] In the same year a suggestion was made that a pier should be constructed out from Government Wharf, which should be large enough to accommodate three steamers. The cost was estimated at £50,000. Jarrett suggests that the scheme was dropped because of objections from local people with financial interests in the lighter traffic.[2] In 1913 a smaller scheme was put forward, but was shelved during the First World War and was never revived. Expansion of the port traffic, both in imports and exports, continued, and following the Second World War it became clear that major improvements were necessary.

Table 5.2

Growth of Freetown's port traffic, 1824–1953
(tons)

	1824	1836	1897	1946	1950	1953
Imports	13,993*	18,372*	26,148	246,796	138,722	187,703
Exports	20,372*	19,901*	11,381	100,330	87,512	129,482
Total	34,365	38,273	37,529	347,126	226,234	317,185

* *Estimate for all Sierra Leone ports.*
  Source: *Jarrett*, Recent Port and Harbour Developments at
  Freetown, *p. 160.*

Queen Elizabeth II Quay was finally opened in 1954. This is a 370-m. (1,203-ft) long deep-water quay capable of accommodating three average ships or two large ones.[3] Cargoes are handled with the help of two 3-ton and one 5-ton mobile electric cranes. Three transit sheds, each with 2,044 sq. m. (22,000 sq. ft) of floor space, provide storage. A warehouse for the Produce Marketing Board was also constructed nearby. The new quay effectively moved the centre of port activity away from Government Wharf. The old port had long suffered from its close proximity to

1 *A. Harley, 'Report on the Freetown Wharf Survey', MS. quoted by H. R. Jarrett,* '*Recent Port and Harbour Developments at Freetown*', Scottish Geographical Magazine, *vol. 71 (1955) 157–64.*
2 *Jarrett, ibid.*
3 *Jarrett, 'The Port and Town of Freetown'. Geography, vol. 40 (1955) 108–18.*

the commercial centre of the city, which meant that little land was available for the expansion of port facilities. The new quay, on the other hand, is some 4 km. (2½ miles) from the city centre, with ample land available on the site and with easy access to the main railway line from the hinterland. It was envisaged that the Government Wharf site would be used for a civic centre, parks and other public facilities, but such development has been slow. This can partly be accounted for by the continued growth of Freetown's port traffic after 1954, for which the newly opened quay soon proved to be inadequate, resulting in the need to continue operations at Government Wharf. A further attempt was made to replace these outdated facilities in 1961, when work started on the construction of a new 120-m. (400-ft) lighter quay to the east of Queen Elizabeth II Quay. This was designed to accommodate five lighters at any state of the tide, and the total cost was some £180,000. This had the effect of increasing the concentration of port facilities in the area to the east of the city around Cline Town. Port traffic continued to increase in the 1960s, continuing the pressure on existing facilities. In July 1964 an agreement was signed with a French company for the construction of a £4½ million extension of Queen Elizabeth II Quay, involving the filling in of much of Fourah Bay. The new quay will have four berths, and is expected to be opened in 1970.

Thus it can be seen that Freetown's port traffic has increased steadily over the years, and the early competitors, such as Bonthe, have been left far behind. However, in recent years Freetown has faced much more severe competition, this time in the export of minerals of various kinds.

Table 5.3

Growth of Freetown's port traffic, 1961–5
(tons)

	Exports	Imports	Total
1961	1,899,141	685,722	2,584,863
1962	2,014,480	695,871	2,710,351
1963	2,118,245	691,008	2,809,253
1964	2,966,888	1,523,385	4,490,273
1965	2,007,668	1,114,376	3,122,044

Source: *Sierra Leone Trade Journal.*

Since 1963, when the chrome mine at Hangha was closed, Freetown has handled no mineral traffic at all. This competition is especially severe when one considers the importance, in terms of value, of minerals to the Sierra Leone export trade (Table 5.4). In 1967 diamonds and iron ore accounted for some 76·6 per cent of total exports by value, and other minerals, particularly bauxite and rutile, are of increasing importance. Special arrangements have been made for the export of each of these (Fig. 5.1).

Exploitation of the iron-ore deposits at Marampa began in 1933. The first ore to be produced came from a residual cap of hard haematite, but

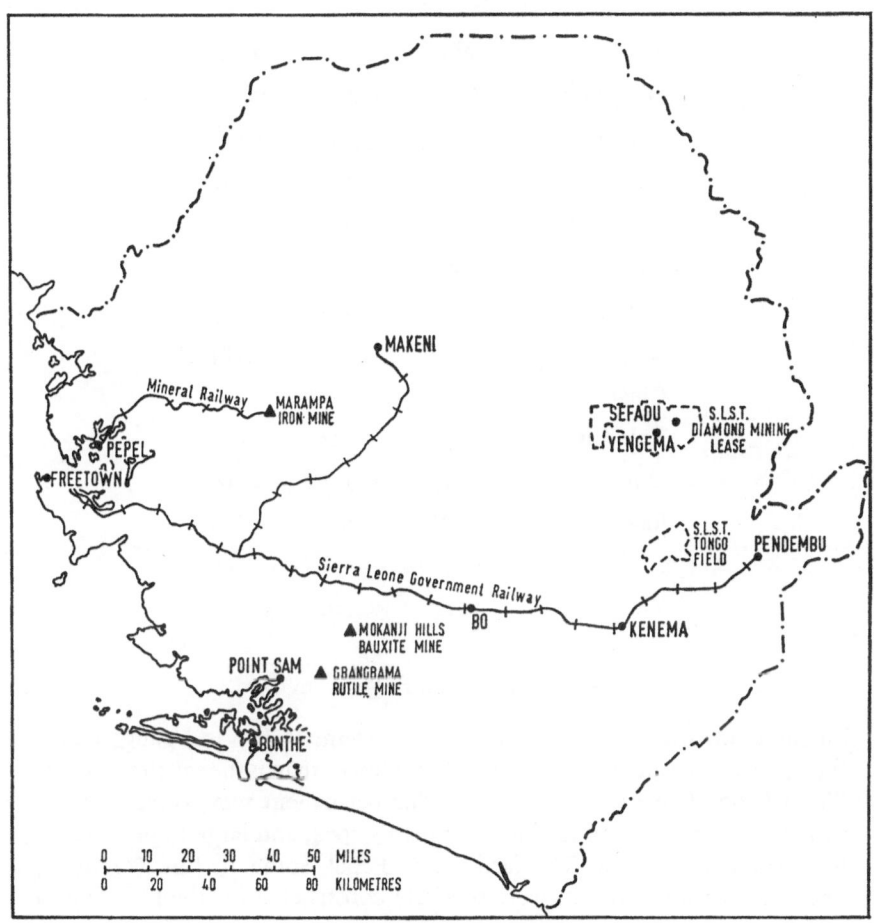

Fig. 5.1   Sierra Leone: main centres of economic activity

this has been completely removed and production is now of the specularite schist forming the main part of the Masaboin Hill at Marampa.[1] The ore has a natural iron content of some 46 per cent, but this is increased to 69 per cent in a new concentration plant opened in 1965. Production of concentrates in 1968 reached 2,455,000 tons, compared with 1,936,000 tons in 1964. The ore is then transported 83 km. (52 miles) to the port of Pepel

Table 5.4

Composition of Sierra Leone's export trade, 1965–7
('000 Leones*)

	1965	*1966*	*1967*
Diamonds	36,959	31,634	29,558
Iron ore	10,896	9,690	9,076
Cocoa beans	903	1,435	1,456
Palm kernels	5,681	5,100	1,099
Bauxite	579	775	929
Piassava	437	168	395
Coffee	1,347	3,923	284
Ginger	321	171	171
Kola nuts	198	241	125
Other	217	321	2,509
Re-exports	5,686	5,672	4,856
Total	63,224	59,130	50,458

* Le.2 = £1

Source: *Central Statistics Office, Freetown.*

on the company's railway, which has a 1·1-m. (3 ft 6 in.) gauge and is therefore more suitable for carrying heavy loads than is the much narrower Sierra Leone Government Railway. The permanent way, originally laid in 1933, was replaced with heavier rails in 1964, and large hopper trucks have been introduced. The facilities at Pepel, some 19 km. (12 miles) upstream from Freetown, were specially constructed for the handling of

1 J. D. Pollett, 'Our Mining Story', Sierra Leone Trade Journal, *vol. 1 (1961) 10–13;*
  K. Swindell, 'Mining', *in* Sierra Leone in Maps, *pp. 92–3.*

the ore (Fig. 5.2). The original loading pier, which was replaced by more modern equipment in 1964, loaded some 31,600,000 tons of ore since its

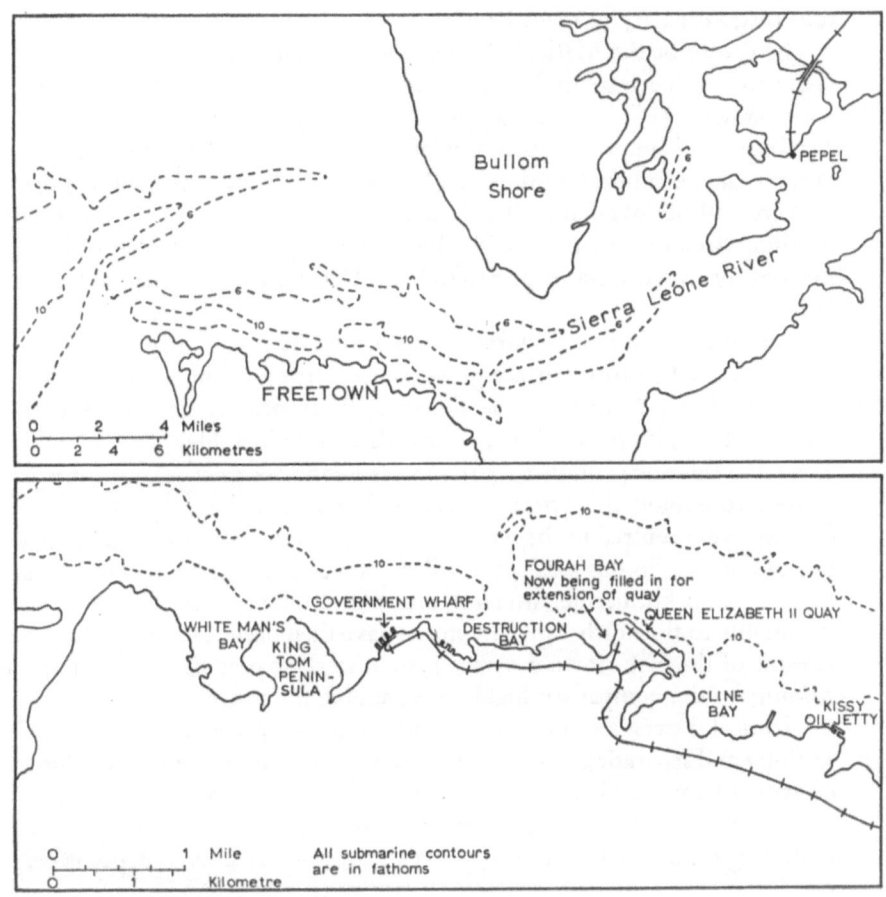

Fig. 5.2   The ports of Freetown and Pepel

opening in 1933. The new Milton Margai Pier, constructed at a cost of £2·5 million, consists of a 720-m. (2,370-ft) jetty supporting a 1·2-m. (4-ft) wide mechanised conveyor. Two 24·4-m. (80-ft) high radial loaders on a 143-m. (470-ft) long pierhead can deliver ore to a waiting ship at a rate of 2,750 tons per hour. The natural channel from Pepel to Freetown was 5·88 m. (19 ft 3 in.) deep, and this was of course completely inadequate for the very large new ore-carriers now in service. At the time of the construction of the new pier the channel was dredged to 9·9 m. (32 ft 6 in.), and has now been further deepened to allow ships of 90,000 tons cargo capacity to enter Pepel.

The future of the Marampa mine, and therefore of the port of Pepel, seems to have been assured for the next ten years by the signing of a large contract with Japan for the supply of ore concentrates. During 1968–9 Marampa will supply 400,000 tons of concentrates, and 1·1 million tons every year thereafter until the end of 1979. It is expected that some £2·5 million will be invested at Pepel, in dredging and the improvement of handling facilities, to cope with this increase. Prior to this agreement most of the iron ore was exported to West Germany, the Netherlands and Britain.

Diamonds were first found in Sierra Leone in 1930, when two gem-stones were taken from terrace gravels near Fortingaira in Kono District. Other finds were made in the following year near Kpara in Kenema District (Fig. 5.1). It soon became clear that a considerable diamond field existed in this area, and in 1935 the Sierra Leone Selection Trust Ltd was formed to exploit this resource. It was shown that the most important deposits were centred to the west of Sefadu, and the company established its headquarters in this area at the village of Yengema.[1] The diamonds are derived from kimberlite, which is found in a series of deep fractures in the granite to the north-east. The stones have then been transported by the streams of the Bafi–Sewa drainage system to their present resting places. Mining by the company is highly mechanised, the gravel being taken by mechanical shovels from the river valleys and then processed in a series of washing and separating plants. In 1962 some 1·8 million cu. m. (2·5 million cu. yds) of overburden were removed and 750,000 tons of gravel were processed.[2] In early 1968 a new diamond retrieval plant was opened, at a cost of £740,000, which can treat over 2,000 tons of alluvial deposits per day.

When mining began, the Sierra Leone Selection Trust was given a concession covering the whole country. During the 1950s illicit hand-mining became important, and caused much concern to the company. Many of these stones were smuggled into neighbouring countries for marketing. In 1956 the concession was renegotiated, giving the S.L.S.T. sole rights in the Yengema and Tongo fields. In a number of other chief-doms licences were given to local diggers, in an attempt to control illicit

1 H. L. *Van der Laan*, The Sierra Leone Diamonds (*London, 1965*).
2 *Swindell, op. cit.*

mining. Marketing was organised through the Government Diamond Office, established in Kenema in 1959. This is now the only legal market for stones from the Alluvial Mining Scheme area. The S.L.S.T. is also obliged to market 50 per cent of its stones here.

Diamonds are, of course, a perfect example of a high-value, low-bulk product, and almost all stones are sent by air to Europe for cutting and polishing, this mode of transport having the advantages of speed and greater security.

Table 5.5

Sierra Leone: rough and uncut diamonds purchased by
Government Diamond Office, 1964–7
(Leones)

1964	23,192,708
1965	22,780,105
1966	19,082,945
1967	21,841,640

Source: Daily Mail, *Freetown*.

Diamonds and iron ore, for so long the backbone of Sierra Leone's export trade, are now being supplemented by the exploitation of the country's deposits of bauxite and rutile. In late 1963 the mining of bauxite began by the Sierra Leone Ore and Metal Company in the Mokanji Hills. The ore is taken 37 km. (23 miles) by road to the specially constructed port at Point Sam, and then by lighter to ore-carriers waiting off Sherbro Island (Fig. 5.1). In 1965 output was 204,000 tons, compared with 148,000 tons in 1964.

Sierra Leone possesses the world's largest deposit of rutile (titanium dioxide), at Gbangbama in the Imperri chiefdom of Bonthe District. Reserves are estimated at 12 million tons. Sherbro Minerals Ltd, a joint British and American company, began production in May 1967, and it was expected that about 100,000 tons would be produced by the end of the year. Initial investment in the project was over £5 million. The ore is transported in covered hoppers over 25½ km. (16 miles) of road to Niti, where it is transferred to 2,000-ton covered barges. These are then towed 29 km. (18 miles) to the Sherbro estuary for loading on to ships.

There are, of course, very good reasons why the various mining companies chose to export their products through completely new ports rather than through Freetown. The special nature of diamonds has already been noted. All the other minerals demanded the construction of completely new handling facilities, and the marginal saving which would have resulted from choosing to build these at Freetown was therefore outweighed by other locational factors. Of prime importance has been the complete unsuitability of the narrow and twisting Sierra Leone Government Railway for mineral transport. If branch lines could have been built from the various mines to the main line, the case for concentration on Freetown would have been much better. As it was, the use of Freetown's harbour would have involved construction of lines all the way from the mines to the capital, and especially in the cases of rutile and bauxite this would have involved the crossing of some very difficult country. This choice did mean the use of sites which were physically far inferior to Freetown, and in the case of Pepel this has necessitated the dredging of the channel on several occasions and at considerable cost.

Freetown and Sierra Leone's economic health

It has become almost a truism amongst writers on development to conclude that the matter under discussion is yet another manifestation of the general ills of economic underdevelopment. Yet the major importance of this in explaining Freetown's relative lack of growth cannot be denied. Sierra Leone, even by the standards of Africa, is a poor country. The fashionable euphemism 'developing', so much despised by the African realists, cannot be applied to Sierra Leone, since it is passing through an economic crisis of major importance. The structural weakness of the economy was clearly shown up by a serious balance of payments problem (Table 5.6), which contributed to the fall of the Government of Sir Albert Margai in March 1967.

Table 5.6

Sierra Leone's balance of payments position, 1964–7
('000 Leones)

	Imports	Exports	Balance
1964	71,019	67,966	− 3,053
1965	76,875	63,224	− 13,651
1966	71,707	59,130	− 12,577
1967	65,268	50,458	− 14,810

Source: *Central Statistics Office, Freetown.*

A balance of payments deficit is not in itself a great disaster, and it could be regarded as an inevitable feature of the early stages of economic growth. However, in Sierra Leone's case the magnitude of the deficit was greater than normal, and a breakdown of the amounts spent on various classes of imports reveals that too high a percentage is accounted for by non-productive goods. Many African countries are forced to increase their import

Table 5.7

Structure of Sierra Leone's import trade, 1965–7
('000 Leones)

	1965	*1966*	*1967*
Food	10,816	13,762	12,458
Beverages and Tobacco	2,622	2,537	2,173
Crude raw materials	969	955	967
Mineral fuels	6,762	5,693	4,818
Oils and fats	1,304	596	1,322
Chemicals	3,914	3,815	3,481
Classified manufactures	19,602	19,197	17,769
Machinery	22,688	16,639	14,803
Misc. manufactures	7,227	7,013	6,350
Misc. transactions	971	1,500	1,127
Total	76,875	71,707	65,268

Source: *Central Statistics Office, Freetown.*

bills by buying machinery of various kinds, but there is a good chance that the investment will increase home produce and thus be profitable in the long term. Between 1965 and 1967 the purchase of machinery by Sierra Leone declined considerably, but non-productive investments in food and manufactured products continued at a high level. The country's export performance has been damaged by the general world-wide decline in prices of primary products, and agricultural produce and iron ore have been particularly hit. This is not the place for a detailed discussion of the structure of the Sierra Leone economy, but Sierra Leone's lack of development must be considered as being of prime importance in explaining the lack of physical port development at Freetown.

An artificially small hinterland

Freetown has the largest natural harbour in West Africa and yet its hinterland, apart from having a low level of economic development, is also extremely small. The competition for colonial territories in West Africa between the European powers resulted in a patchwork of small states. Each of these colonies was administered purely with the needs of the metropolitan power in mind. Agricultural and mineral production were destined for the market of the ruling power, and the communications pattern was designed with this exploitation in mind. The roads and railways of neighbouring countries were generally not linked, hence contact of any kind was at a minimum. Such was the case with Sierra Leone and its French-ruled neighbour Guinea. Each power built a railway (of different gauge) to the interior of the country and all trade was channelled to the respective capital. As Sierra Leone is surrounded on the north and east by Guinean territory, this had the effect of restricting Freetown's effective hinterland to the 73,326 sq. km. (27,925 sq. miles) of Sierra Leone itself. This small area has considerable mineral resources, but its agriculture is poorly developed. Although Sierra Leone is one of the more densely populated countries of Africa, with 30 people per sq. km. (78 per sq. mile) in 1963, the total population of 2,180,355 (1963) offers a very limited market, especially when the extremely low per capita purchasing power is considered.

Without the extremely effective trade barriers erected by the colonial powers, Freetown could well have been expected to take in much of the eastern part of Guinea and perhaps parts of Mali to its hinterland. However, the communication patterns which have developed have so far effectively ruled this out.

An avoidable waste of natural potential?

We have looked at the relative lack of development of Freetown's magnificent natural harbour, and the factors which can be put forward to explain this are the small and poorly developed hinterland which Sierra Leone provides, and the diversion of the most valuable part of the export traffic to other ports. We can now ask what are the prospects of more use being made of the harbour, thereby creating a demand for additional port facilities.

Before trying to answer this question, we should consider first the desirability of such development taking place. The growth of the port cannot be considered in isolation from the expansion of the city, and what one is really asking is whether Freetown should grow any larger, at least for the time being. Although population data before 1963 are no more than estimates, it seems that Freetown is growing at a rate of just under 10 per cent per year. The 1963 census showed that 5·9 per cent of the total population of Sierra Leone live within the Freetown city area. If one includes the suburban settlements outside the city boundary but really forming part of the urban area, the percentage rises to 7·5.[1] Neither the growth rate nor the level of concentration in the capital city are exceptional. A number of African cities are experiencing growth rates of 15 per cent per year, and in many countries dominance of a single urban centre is more marked. For instance, over 15 per cent of the population of Senegal live in the city of Dakar. Sierra Leone should be thankful for these low figures. The growth of many African capitals is so fast that the provision of either jobs or houses for many of the newcomers is completely beyond the resources of the authorities concerned. At the same time those urban dwellers who do become established in a job are quickly able to enjoy a standard of living far above that found in the rural areas. This mal-distribution of income naturally encourages more people from the up-country areas to search for their fortunes in the capital. The pattern of roads and railways established by the colonial powers assured primacy to the port capitals of West Africa, and it would seem that most of these are over-large for the economic and social well-being of the countries as a whole.

This is not to deny the need for modern and efficient ports in West Africa, and in most cases the very real economies of scale in port opera-tions ensures that efficiency and large size go together. The need for a modern port is especially acute when it is remembered that the Sierra Leone economy depends on the export of primary raw materials, many of which are extremely bulky. Green and Seidman[2] have stressed the dangers of the dependence on exports common to many African countries, yet in the case of Sierra Leone such dependence on primary products is unfortunately bound to continue for the foreseeable future, and with it the need for the kinds of exporting facilities existing at present. The short-term need to overcome the present balance of payments problem will

1 J. I. Clarke, 'Population Distribution in Sierra Leone', in J. Caldwell and C. Okonjo (eds.), Population of Tropical Africa (London, 1968) pp. 270–7.
2 R. Green and A. Seidman, Unity or Poverty (London, 1968) chap. 2.

ensure the continuation of this situation, although efforts are being made
to find a long-term solution to the problems of development. Of
particular importance are attempts to modernise agriculture, and thereby
reduce the need to import food.

The important question to ask about the desirability of increased port
development at Freetown is whether, when increased development takes
place in the country as a whole, the port's physical and other advantages
will enable it to provide the necessary services at a real cost to the whole
community which is low enough to offset the dangers of an overgrown
capital city already mentioned. As has been stressed, these physical
advantages, allied with the established pattern of communications, are
considerable. The recent decision to close the Sierra Leone Government
Railway, while depriving Freetown of a major hinterland link, should not
seriously damage this dominant flow. Also, the present population of the
Freetown area provides a relatively large market which, along with the
advantage of being close to a major port, will encourage entrepreneurs to
establish their industrial plants in the city. The multiplier effect in such
a case will be powerful. However, the decision to establish special ports
for mineral exports does show that these forces are not all-powerful, and
the community growing up at Pepel, for example, does demonstrate the
possibilities of decentralisation and regional economic development that
are in the real interest of the country.

Laying aside the question of whether the port of Freetown really ought
to expand, it seems likely that it is only likely to grow if it is able to
expand its effective hinterland, or if Sierra Leone becomes an economically
more prosperous hinterland. Many would argue that the only real possibility of either of these conditions being fulfilled is through the unity of
the numerous small states in the area. The economic cases for such a
belief has been convincingly stated in the recent study by Green and
Seidman. The difficulties facing any attempt to implement such proposals
are of course immense. Sierra Leone and her neighbours have contrasting
languages, cultures and political creeds. The recent coups in Sierra Leone
also ensure that her government will be primarily concerned with
internal affairs at least for the time being. This probably explains the
country's absence from the recent meeting in Monrovia to set up a West
African Regional Group, designed to increase economic co-operation. At
the time of writing it is not clear how effective this new grouping will be,
or what will be Sierra Leone's relationship with it.

Over the last few years there have in fact been a number of attempts in
West Africa to create regional groupings, but so far without any real

success. In 1964 there was an attempt to set up a free-trade area between Sierra Leone, Guinea, Liberia and Ivory Coast. This proposed grouping had the advantages of geographical contiguity, but the political difficulties involved in trying to marry two states so completely different as Guinea and Liberia proved to be too great. The high hopes at that time, and the plans for development that were put forward, illustrated the possibilities for such a grouping. In particular the wastefulness of duplicating expensive plant such as steel mills and oil refineries was stressed. One shared plant of this kind could easily supply the total market in that part of the continent. As an obvious site for an oil refinery and other industries, the port of Freetown would have benefited from such development. There were also plans to rationalise the patterns of communications and trade, which would probably have resulted in the trade of part of Guinea passing through Freetown. In fact Freetown now has an oil refinery to serve the national market, but all the other developments are still only pious hopes. The benefits that would result from such unity are now clear to all, but as yet each country seems to feel that these are outweighed by the political and other changes in their lives that would result. Until such co-operation takes place there seems to be no hope of the major economic breakthrough that Sierra Leone needs, and meanwhile the physical potential of Freetown's harbour must await its full development.

6. The Ports of Liberia: Economic Significance and Development Problems

W. Schulze

Professor Dr Willi Schulze took his Ph.D. degree at Marburg University in 1950, and subsequently taught geography in South America for five years. From 1962–6 he was Lecturer in Geography in the University of Liberia, and is the author of a volume on the geography of Liberia. Dr Schulze is currently on the staff of the University of Giessen.

The rapid economic development of Liberia since 1945 has been closely connected with the development of modern ocean terminals. When in 1934 the rich magnetite deposit of the Bomi Hills (Fig. 6.1) was discovered by the Dutch geologist Terpstra, the syndicate of the William H. Mueller Company could not make use of its concession because of the absence of suitable port facilities. In 1938 geologists from the United States Steel Company investigated the ore body and made an unfavourable report in which 'the difficulty of shipping such a material as iron ore by surfboat operation' was emphasised, with the result that the company lost interest.[1] It was only in 1946, when the construction of the free port of Monrovia was well advanced, that the Liberia Mining Company was established by the American businessman and engineer Landsdell K. Christie. With the completion of the port in 1948 and the first shipment of iron ore from the Bomi Hills in 1951, a new chapter in the economic history of Liberia began, during which the country has become not only the main producer of the mineral in Africa but also one of the principal suppliers of the world market. Without her modern ports Liberia could not have reached her present level of economic development.

Geographical and technical characteristics of the Liberian ports

The economic significance and the problems of development of any port depend on a variety of geographical factors, physical and human, combined with a range of technical considerations. These have been

1 R. E. *Anderson*, Liberia, America's African Friend (*Chapel Hill, N.C., 1952*) p. *181*.

summarised for the Liberian ports in Tables 6.1 and 6.2. Physical geographical factors strongly influence the efficiency of a port. Oceanographic features such as water depths, seaboard obstacles, tides and waves may influence

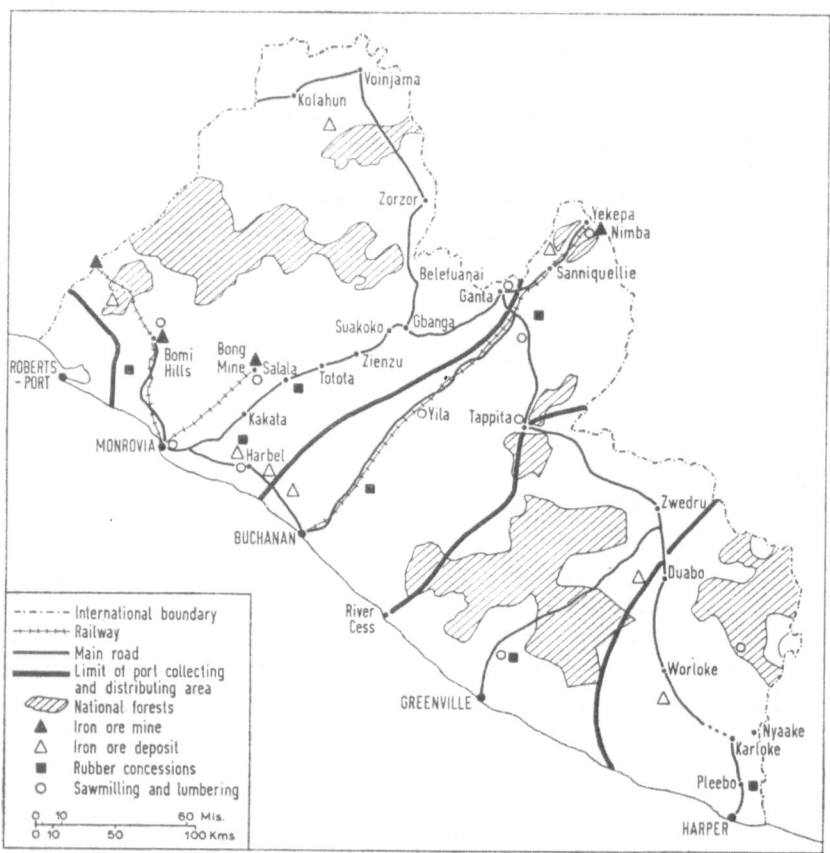

Fig. 6.1　Economic map of Liberia

the accessibility and the use of port facilities. Climatic elements such as temperature, rainfall and humidity may interfere with the loading and landing operations or necessitate special storage or cooling facilities. Soil conditions may create problems for the building of heavy structures or for the stockpiling of heavy loads, e.g. Liberian iron ore. Almost of greater significance is the geographical character of the hinterland or ports' collecting and distributing area (P.C.D.A.). The location and development of ports is closely related to the volume and distribution of primary and secondary production and consumption, which in turn reflects the number, density, distribution, occupation, structure, and degree of urbanisation of the population. The association of port and P.C.D.A. depends on an effective network of roads and railways, and transportation and economic structure are closely dependent.

Monrovia emerges as the port with by far the best-developed P.C.D.A. (Table 6.1). In area (45,850 sq. km., or 17,500 sq. miles) it is larger than those of Robertsport, Greenville and Harper together, and as the location of the country's original deep-water port and first highway (Monrovia–Gbanga in 1945), it has attracted most of the large foreign enterprises – mining, agricultural and manufacturing. Two-thirds (2,450 km., or 1,530 miles) of the country's total road network of 3,680 km. (2,300 miles) are found in the western part of Liberia. The average road density is 5·3 km. per 100 sq. km. (8·7 miles per 100 sq. miles), which is nearly four times that of the Greenville P.C.D.A. Monrovia's road links include nearly all the country's paved roads and also two of the international connections, namely, via Ganta and N'Zérékoré to Guinea and via Foya and Kolahun to Sierra Leone. In addition Monrovia serves as the coastal terminus for two of the country's three railway lines. Over half of Liberia's population lives in the area served by Monrovia, and the population density of 13·5 persons per sq. km. (35·4 per sq. mile) contrasts particularly markedly with the population densities of Greenville and Harper. Monrovia itself has a population of 110,000 and there are major concentrations in Montserrado County and along the main highways. Along the Kakata–Ganta section are fifteen clan areas totalling 84,000 inhabitants, to which may be added the 50,000 persons on the Firestone plantation between Kakata and Harbel. Given local conditions, a settlement of 1,000 inhabitants may be considered urban and Monrovia has in its hinterland twenty-three of the country's fifty towns. Altogether some 70 per cent of Liberia's employed persons (70,000 out of 100,000) are in Monrovia's P.C.D.A.

Buchanan's hinterland is clearly less well developed. The overall population density is marginally lower and a reasonable degree of urbanisation is reflected in a relatively large number of employed persons. The economic structure is in fact dominated by the LAMCO mining operations at Mount Nimba, the abundance of timber with two important operating companies (MIM and SIGA), by a number of Liberian rubber farms, and by two rubber concessions of which the Liberian Agricultural Company's (UNIROYAL'S) 7,200 hectares (18,000 acres) is the second largest in the

Table 6.1
Geographical characteristics of Liberian ports and their hinterlands (P.C.D.A.s)

Location and Physical geography	Robertsport	Monrovia	Buchanan	Greenville	Harper
Geographical position	6° 46′ N 11° 30′ W	6° 21′ N 10° 48′ W	5° 52′ 12″ N 10° 03′ 49″ W	4° 59′ 30″ N 9° 02′ 00″ W	4° 22′ 01″ N 7° 43′ 43″ W
P.C.D.A. (sq. miles)	600	17,500	9,450	8,700	6,650
Rainfall (in.)					
Annual average	169·86	182·58	158·06	173·84	119·95
Average maximum (Month)	36·74 (July)	36·34 (June)	31·14 (September)	32·77 (June)	21·38 (May)
Average minimum (Month)	1·17 (Jan.)	1·77 (Jan.)	1·25 (Jan.)	3·81 (Jan.)	3·79 (Aug.)
Temperature (° F.) Annual average	78·8	78·9	78·8	75·8	76·9
Water temperature Av. min./av. max.	78·9/82·5	78·9/82·5	77·8/82·5	77·0/82·5	76·2/82·5
Maximum tidal range (ft)	4·0	4·0	4·3	4·5	4·5
Economic structure					
National forests (million acres)	—	1·6	0·2	1·75	0·65
Iron-ore mines	—	3	1	—	—
Production 1967 (million long tons)	—	9·5	8·1	—	—
Rubber concessions	—	3	2	1	1
Rubber farms (est.)	few	3,300	200	75	650
Production, 1966 (est.)	—	47,000 tons	1,000 tons	1,200 tons	5,500 tons
Rubber area, 1966 (est.)	100 acr.	223,000 acr.	26,000 acr.	7,000 acr.	25,000 acr.
Lumbering (companies)	—	4	5	1	1
Electric energy (kW.) (capacity, est.)	150	110,000	50,000	1,000	1,000
Infrastructure					
Road length (miles)	60	1,530	340	210	180
Paved	—	(170)	(40)	—	—
Miles per 100 sq. miles	10·0	8·7	3·6	2·4	2·7
Railways (miles)	—	141	167	—	—
Miles per 100 sq. miles	—	0·8	1·9	—	—
Population structure					
Total population (est.)	20,000	620,000	256,000	48,000	74,000
Average density per sq. mile	33·3	35·4	27·1	5·5	11·1
Persons in money econ.	1,000	70,000	18,000	4,000	7,000
Urban centres	1	23	21	2	3
Inhabitants of port city 1968 (est.)	3,000	110,000	14,000	4,000	7,000
Census 1962	(2,417)	(80,992)	(11,909)	(3,962)	(6,095)

country after Firestone's Harbel plantation of 30,000 hectares (75,000 acres).

The Greenville hinterland has only one rubber concession but does contain the principal forest reserves of the country, the Krahn-Bassa National Forest. A new road from Buchanan to the Zwedru–Ganta highway was opened in 1967 and has greatly improved the structure of the transport system, but even now there is still only 1·4 km. of road for every 100 sq. km. (2·4 miles per 100 sq. miles) and the population density is only 2·1 per sq. km. (5·5 per sq. mile). The new road will make a potential 105,000 tons of timber available for export every year, and shipments of at least 30,000 tons from Greenville will mean a new phase of port development.

Like the Greenville P.C.D.A., that of Harper has a low population and route density, only one large rubber concession (Firestone's Cavalla plantation of 13,000 acres), but large forest resources. Robertsport is the smallest of the hinterland areas and does not extend north of the new road from Clay (Kle) to Puyehun which tends to channel traffic to Monrovia. The small P.C.D.A. has a relatively high population and road density, but the absence of mining and commercial agriculture on any scale precludes great port activity.

The technical characteristics of the Liberian ports have been summarised in Table 6.2. Monrovia has by far the largest harbour, and the two dolerite (diabase) block breakwaters enclose a triangular basin of 300 hectares (750 acres) (Fig. 6.2; Plate 3). An H-pile marginal quay of 610 m. (2,000 ft) acts as a commercial quay, and pipeline loading of latex is possible at the southern berth which is connected to two 500-ton-capacity storage tanks. A petroleum berth has been provided (1962) as a jetty extending northwards from the southern breakwater, and the northern breakwater has been equipped with a fishing pier (1963) and cold store (1964). Iron ore is handled at three finger quays built between 1962 and 1965 by the Liberia Mining Company (L.M.C.), the National Iron Ore Company (NIOC) and the German Liberian Mining Company (DELIMCO). The port has a land area of 205 hectares (507 acres), most of which is leased by trading companies (twelve), oil companies (four), mining companies (three) and rubber companies (three). The three mining companies take over 85 hectares (210 acres), and L.M.C., NIOC and DELIMCO have stockpiles of 460,000, 250,000 and 500,000 tons respectively. The oil-tank farms have 40,000 tons capacity and are being extended to serve an oil refinery recently established at the Paynesville industrial estate.

Table 6.2

Technical characteristics of Liberian ports

Specification	Monrovia	Buchanan	Greenville	Harper
Construction				
Construction period	May 1944 –July 1948	Jan. 1961 –July 1963	1955–9, 1962–4	1958–9
Construction costs (million U.S. $)	19·43	20·53	7·70	1·60
Harbour basin				
Area of basin (acres)	750	167	58	10
Main breakwater (ft)	6,560	6,200	1,270	500
Secondary breakwater (ft)	6,560	2,000	—	350
Width of entry (ft)	853	700	1,220	320
Maximum draught (ft)	Ore piers 38 Commercial quay 30	Ore quay 42 Commercial quay 33	21	18 at inner pier
Cargo facilities				
Commercial quay, length	2,000 ft	1,095 ft	590 ft	180 ft
Commercial quay, width	30·6 ft	240 ft	45 ft	20 ft
Cranes (no./capacity)	1/70 ton, 1/25 ton 2/10 ton, 1/5 ton	1/150 ton 1/20 ton	1/10 ton —	1 truck crane (priv.)
Transit sheds (no./sq. ft)	3/104,700	—	1/6,500	1/5,000
Warehouses	5/60,300	—	1/10,000	1/5,000
Marine equipment				
Tugboats (no./h.p.)	2/400, 1/280 1/186	2/1,640	1/800	—
Barges, boats, etc.	1 barge 10 lighters	—	surf boats	surf boats
Navigation aids				
Range lights	2	2	1	—
Radio telephone	V.H.F.	V.H.F.	V.H.F. (tug)	—
Special facilities				
Iron-ore loading	3 finger piers 855–910 ft long	ore quay 843 ft long	—	—
Ore-loading rate (Max. per hour)	1,200–2,000 tons	6,000 tons	—	—
Fishery	Fishery piers	—	—	—
Cold storage	2	—	—	—
Latex handling	Tanks, 2,420 tons	—	—	—
Port operations (1966)				
Number of calls	1,641	319†	79	120*
Tonnage landed (long tons)	601,606	88,600†	11,546	16,132‡
Tonnage loaded (long tons)	8,573,903	7,985,035†	2,687	10,244‡
Tonnage landed and loaded	9,175,509	8,073,635†	14,461	26,376‡
Iron ore shipped (long tons)	8,433,764	7,979,035	—	—
Total value of exports (U.S.$)	90,000,000†	53,000,000†	800,000†	2,800,000†

* 1962.
† Estimates.
‡ Fiscal year 1966–7.

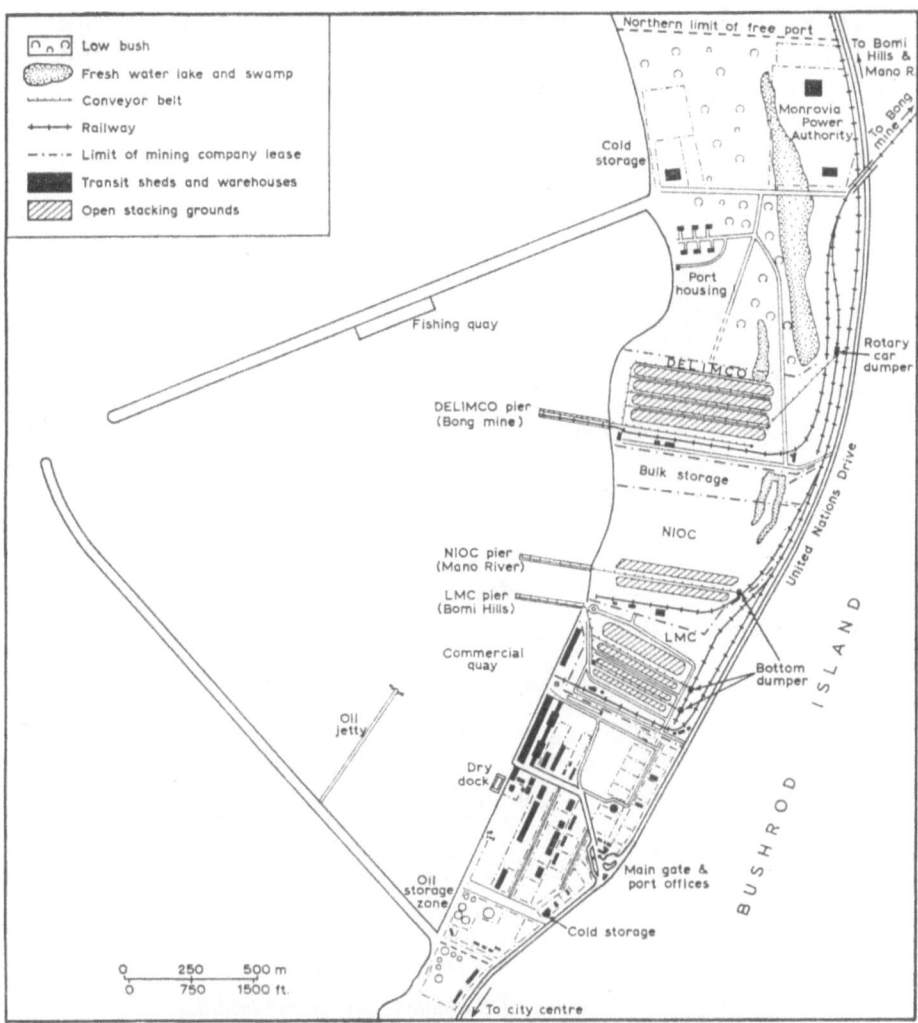

Fig. 6.2   The port of Monrovia: layout and facilities

The harbour at Buchanan (Fig. 6.3; Plate 4) was built for the Liberian American Swedish Minerals Company (LAMCO) and Bethlehem Steel for the export of iron ore mined at Nimba. Breakwaters enclose a basin of 67 hectares (167 acres), and the ore-loading quay with 12·8 m. (42 ft) alongside is located opposite the main breakwater. Vessels of 45,000 tons deadweight can be accommodated at all times and ships of 70,000 tons can

D

be berthed by using the 1·3-m. (4·3-ft) tide. The ore-loading facilities have a rated capacity of 6,000 tons an hour and areas have been set aside for ore stockpiles and processing. All the mineral now passes through a

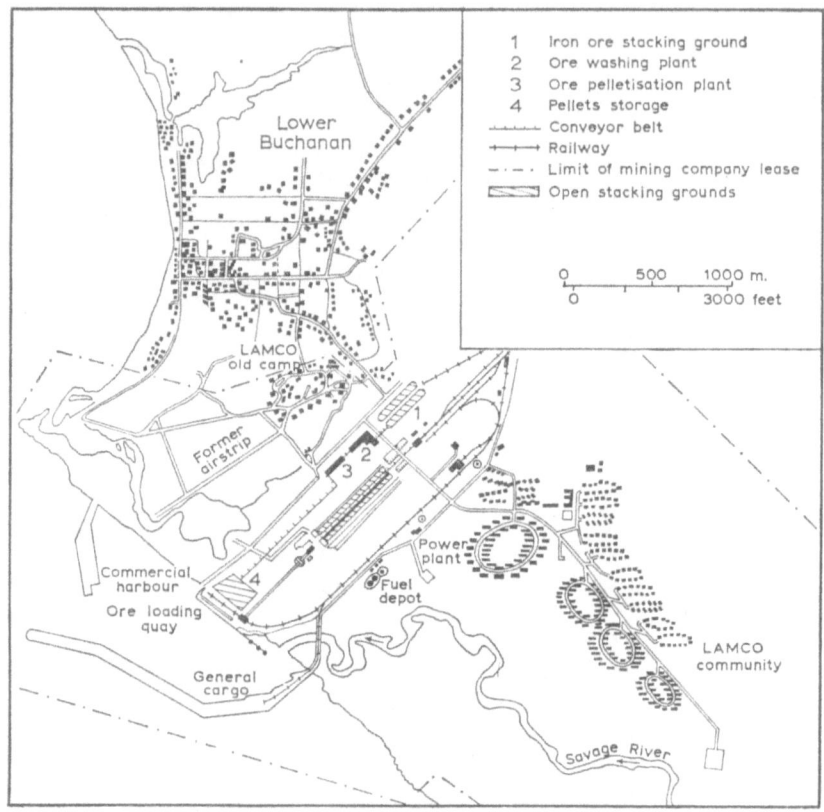

Fig. 6.3　The port of Buchanan: layout and facilities

washing plant, and as much as 2 million tons a year is now being shipped as pellets.

The facilities available at Greenville (Fig. 6.4) and Harper are limited. The areas of protected water are small, the quays of no great length and shed accommodation restricted. Harper does not even have its own cranes.

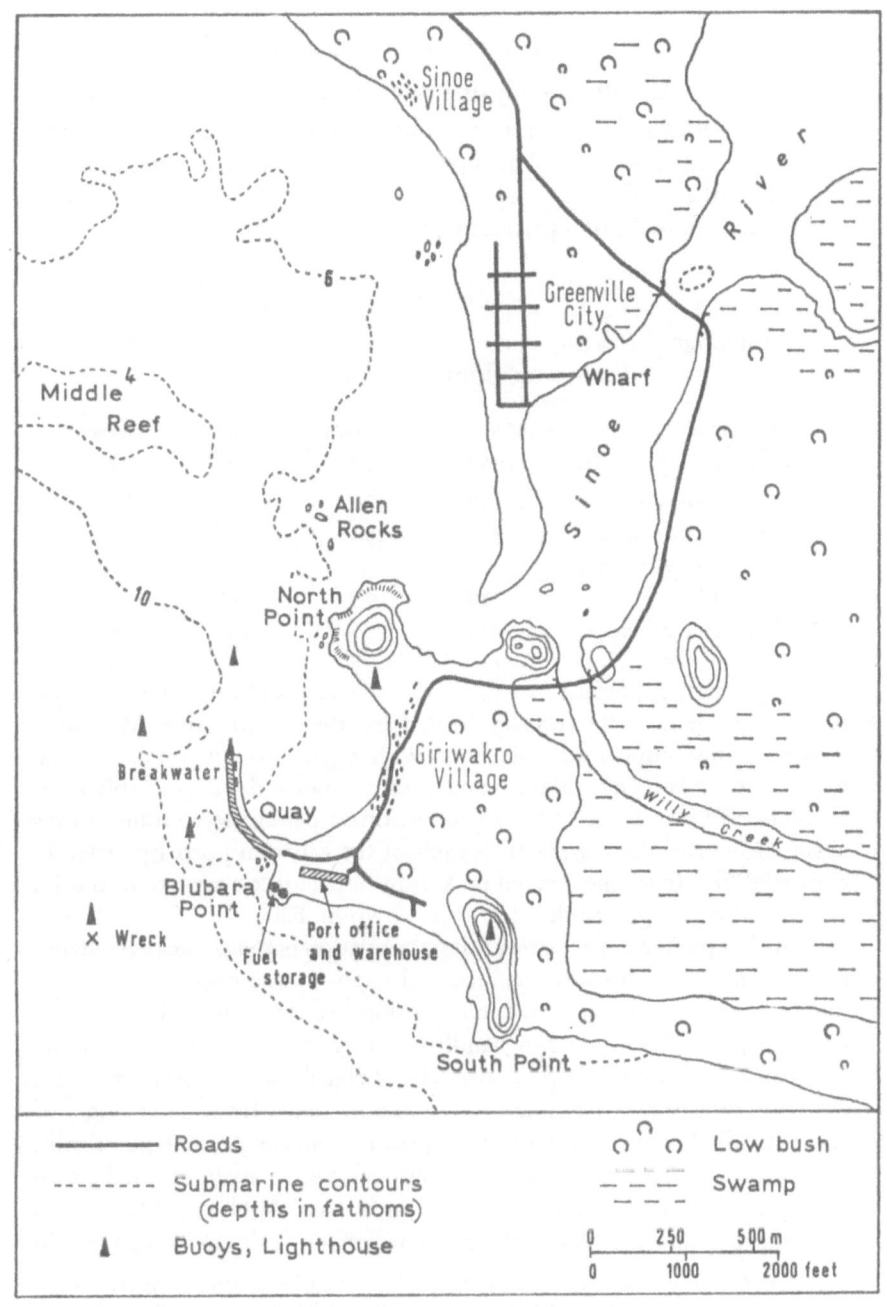

Legend

────────	Roads	o o o o o	Low bush
----------	Submarine contours (depths in fathoms)	─ ─ ─	Swamp
▲	Buoys, Lighthouse		

0		250	500 m
0	1000		2000 feet

Fig. 6.4   The port of Greenville

Economic significance of the ports

The economic significance of the free port of Monrovia is indicated by the figures compiled in the 'Port operations' section of Table 6.2.[1] Its position as the main seaport of the country becomes even clearer when the absolute figures are replaced by percentages (Table 6.3).

Table 6.3

Ranking of Liberian ports by percentage share of total traffic
(based on figures in Table 6.2)

Port operations	Monrovia	Buchanan	Greenville	Harper
Total number of calls	76·2	14·8	3·4	5·6
Total tonnage landed	84·0	12·4	1·4	2·2
Total tonnage loaded	51·7	48·2	0·01	0·06
Total tonnage landed and loaded	53·1	46·7	0·07	0·15
Total value of exports	61·4	36·2	0·5	1·9

The general importance of the port is clearly indicated by the high percentages of calls, the tonnage landed and the export value. Monrovia is the leading import harbour, but with regard to the total tonnage loaded the advantage over Buchanan is only small and will probably even be reversed when the LAMCO production surpasses the 10 million tons level. However, the comparative rank of the Liberian ports by order of exports differs from the general rank with respect to certain commodities and has also changed with time, as is shown in Table 6.4.

A further point emerges concerning the relation between the percentages of calls and those of cargo handled and export value respectively. In an ideal case the percentage of a country's ships' calls to a port should equal its percentage share of cargo handled and of export value, so that the quotient would be 1:1, e.g. 50 per cent of ships' calls would correspond to 50 per cent of cargo handled and of export value. If the percentages of cargo handled or of export value are greater than the percentage of calls, the port concerned has a positive quotient which may be used as an additional indicator of its economic significance. Applying this method to Monrovia and Buchanan, the relation between calls and cargo would

1 *Some figures are marked as estimates or are based on fiscal years. Regional data are not easily available in Liberia and sources often give varying figures, as is also noted by other authors, e.g. W. A. Hance in* African Economic Development, *2nd ed. (New York, 1967) p. 61.*

Table 6.4

Ranking of Liberian ports in order of exports
((a) = 1950–63; (b) = 1966)

Commodity	Monrovia		Buchanan		Robertsport		River Cess		Greenville		Harper	
	(a)	(b)	(a)	(b)	(a)	(b)	(a)	(b)	(a)	(b)	(a)	(b)
Iron ore	1	1	2	2	–	–	–	–	–	–	–	–
Rubber	1*	1*	–	4	–	–	–	–	3	3	2	2
Palm kernels	1	1	2	2	6	6	3	3	5	5	4	4
Coffee	1	1	2	3	4	4	5	5	6	6	3	2
Cocoa	1	1	5	5	6	6	4	4	3	3	1	2

* *Transhipments from Harbel via Farmington river.*

Source: *Adapted from Stanley*, Final Report, *p. 18, with ranks for 1966 added.*

be 76·2 : 53·1 and 14·8 : 46·7 respectively, i.e. the quotient would be 0·70 for Monrovia and 3·16 for Buchanan. As to the call/value quotient, the figures would be 0·81 for Monrovia and 2·56 for Buchanan. In both cases Buchanan ranks considerably higher than Monrovia, particularly in the call/cargo quotient.

The total cargo handled at Monrovia increased seventy-five-fold from 1950 to 1967, from 133,578 tons to 10,076,000 tons. This rapid development was mainly due to the advent of the iron-ore industry, which started exporting in 1951 (170,707 tons) and shipped over 9 million tons in 1967. The growing amount of general cargo handled and of fuel imports contributed to the growth of port traffic, as shown in Table 6.5.

Fig. 6.5 shows a general increase of freight movement, but the growth varied for commodities and with time. Iron-ore exports recorded the fastest growth, petroleum imports increased more than five times from 1952 to 1967, while the tonnage of general cargo handled was only 3·7 times greater in 1967 than in 1950. The smallest increase was in the number of ships' calls. Decreases of general cargo handled in 1964 and 1967 and of fuel landed in 1967 are both due partly to periods of economic recession and partly caused by the completion of large building projects at the mines, at the Mount Coffee hydro-electric plant and on the road network.

The number of ships calling at the port of Monrovia increased from 1,154 in 1958 to 1,622 in 1967 (Table 6.6). The greatest increase was in

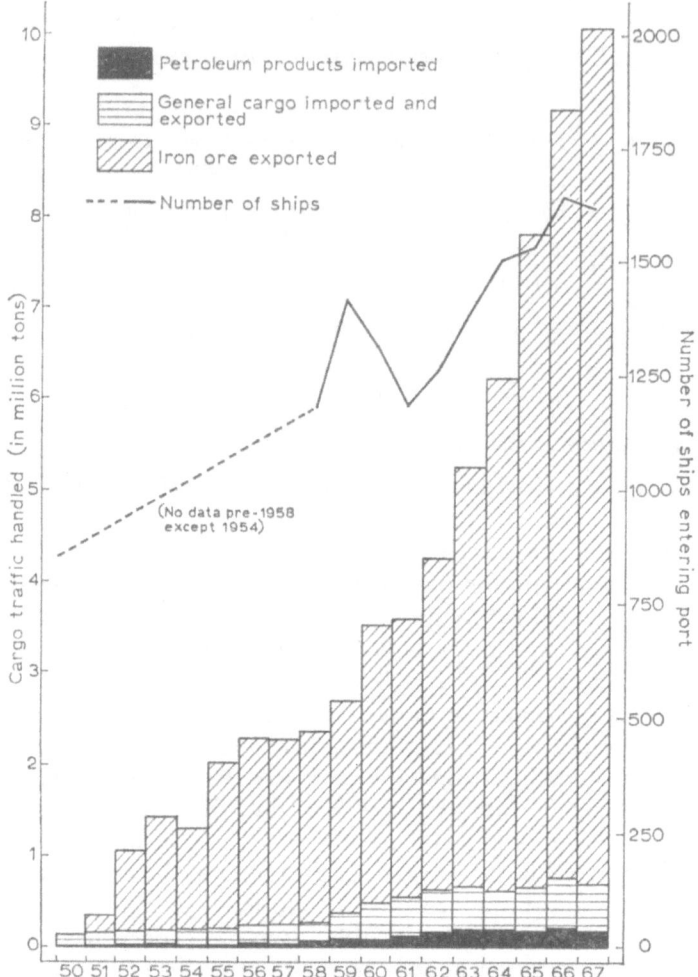

Fig. 6.5   Cargo traffic at the free port of Monrovia, 1950–67

bulk cargo carriers, mainly of iron ore, the number of which more than doubled within the ten years. During the same period the number of coastal vessels remained nearly constant.

The coastwise service is carried out by Farrell Lines, which opened a feeder service along the coast in December 1947 and made an agreement with the Firestone Plantations Company in 1950 concerning the transport of rubber latex to Monrovia from the Farmington river via Marshall, and

Table 6.5
Cargo movements at the free port of Monrovia, 1950–67
(long tons)

Year	General cargo landed and loaded	Iron ore loaded	Petroleum landed	Total cargo handled
1950	133,578	—	—	133,578
1952	142,988	888,140	34,446	1,065,574
1954	164,746	1,190,051	23,951	1,278,748
1956	212,624	2,037,449	40,042	2,290,115
1958	241,073	2,045,375	52,806	2,339,254
1960	396,156	3,037,540	81,025	3,514,721
1962	478,854	3,642,984	137,796	4,259,634
1964	445,748	5,649,476	161,025	6,256,249
1965	490,145	7,152,760	169,740	7,812,645
1966	517,199	8,433,764	224,546	9,175,509
1967	503,348	9,394,448	178,029	10,075,825

Source: *Brown Engineers*, Transportation Survey, *Annex 2*; B. Naidoo, Liberia Annual Review, 1960–1961 (*London and Monrovia, 1960*) *p. 65; cargo statistics provided by Monrovia port management.*

Table 6.6
Number of ships calling at the free port
of Monrovia, 1958–67

Year	General Cargo Main-line	Coastwise	Bulk cargo carriers	Total
1958	529	449	175	1,154
1960	629	455	235	1,319
1962	581	420	248	1,249
1964	768	435	302	1,505
1965	758	422	348	1,528
1966	823	452	366	1,641
1967	777	446	399	1,622

Source: *As for Table 6.5.*

from Harper.[1] The coastwise cargo handled in the harbour of Monrovia decreased from 17,778 tons in 1964 to 10,057 in 1967. During the same four years the volume of transhipments carried out in the free port grew from 58,000 tons to over 85,000 tons, while the general cargo loaded stagnated between 35,000 and 45,000 tons per year.

The port of Buchanan has a large potential collecting and distributing area of 24,660 sq. km. (9,450 sq. miles), containing one-quarter of the country's total population. Unfortunately most of the people live in Nimba County, some 240 km. (150 miles) from the coast. Although according to the map this area has direct road and railway connection to the port, in reality the transportation problem is unsolved for two reasons. Firstly, the LAMCO road from Buchanan to Gba (340 km., or 130 miles) was built as a service road to facilitate the construction of the Nimba–Buchanan railroad. In 1963 maintenance was left to the Liberian Government, which had neither the personnel nor the technical or economical means to do the job properly. As a result only certain sections of the road, including a portion in the south which is taken care of by the Liberian Agricultural Company (L.A.C.), is usable all the year round. Secondly, until 1966, the use of the railway was restricted exclusively to the transport of iron ore, except for a railbus for LAMCO personnel. In 1967 the first shipment of logs was made from Buchanan. There are two timber companies, near Yila (SIGA Lumber Company) and near Tappita (MIM Timbers of Liberia), which formerly had to ship their logs or sawn timber to Monrovia by truck via Ganta, over 369·5 km. (231 miles) (SIGA) and 426 km. (266 miles) (MIM) respectively. Transportation costs, including the return of empty trucks, amounted to U.S. $16 per cu. metre (about a ton), and can be reduced to U.S. $11 by utilising the LAMCO railroad.[2] By operating via Buchanan instead of Monrovia, the freight rate from Yila to the port will be reduced to 5·15 cents per ton-mile instead of an average rate of 12·0 cents.[3] As a result, a considerable increase of timber exports via Buchanan can be anticipated.[4]

Other factors will contribute to augment the economic significance of the new port. In 1967 a siding was built at the L.A.C. plantation, and in

1 R. G. *Albion*, Seaports South of Sahara (*New York, 1959*) *p. 232.*
2 *W. R.* Stanley, Final Report on Selected Economic and Social Benefits to be Derived from the Construction of Nine Separate Lengths of Rural Access Road in Liberia (*Washington, D.C., 1967*) *p. 20.*
3 *Republic of Liberia, Office of National Planning,* Using the LAMCO Railway for Timber Exports (*Monrovia, 1965*) *p. 2 (mimeographed).*
4 *A first shipment of 4,000 tons of timber was made from Tropoi, south of Gba (179 km. (112 rail miles) from Buchanan), on 26 March 1967.*

January 1968 the first 270 tons of rubber were shipped by railroad. In addition, the Cocopa plantation of the Liberia Company near Flumpa to the south-east of Ganta is interested in using the rail connection. Another customer will be the Republic of Guinea, which formerly trucked some 15,000 tons of cargo annually to and from the free port of Monrovia, using the principal highway via Ganta (211 km., or 182 miles). In 1967 a new agreement was signed between the two governments concerning the utilisation of the LAMCO line for transit trade.[1] The distance from Ganta to the coast will be reduced to 200 km. (125 miles), and at the same time the deleterious effects on the road and on the traffic in Monrovia city and port will be eliminated. Petroleum products, cement and other building materials as well as agricultural produce will also utilise the railway and the commercial quay at Buchanan and will increase the economic significance of the new port. At present, the port is little more than a mineral port, and 98·9 per cent of the total tonnage handled in 1966 consisted of iron ore. Iron-ore shipments started on 14 July 1963 and in that year reached 1,797,000 metric tons, which were transported by some 85 ore-carriers. In 1964 exports reached nearly 7 million tons, and from 1965 to 1967 the figure stayed constant near 8 million tons, as shown in Table 6.7.

Table 6.7

Iron-ore shipped and number of ore-carriers
calling at the port of Buchanan, 1964–7
(metric tons)

Year	Tons of iron ore shipped	Number of ore-carriers
1964	6,875,124	215
1965	8,316,310	244
1966	8,106,790	204
1967	7,759,350	186

Source: Goeransson, *The Port of Buchanan, Diagram 1, and information by LAMCO J.V., Liberia.*

In January 1968 Africa's first pelletising plant with a capacity of 2 million tons a year, and a 10-million-ton washing plant, were put into

[1] *Republic of Liberia, Department of Planning and Economic Affairs,* Annual Report, 1966–1967 *(Monrovia, 1967) p. 71.*

operation at Buchanan and will contribute to the further increase of LAMCO's ore exports. In the year 1968 exports amounted to 7·6 million tons of ore and 1·1 million tons of pellets. Shipment of the ore is mainly to Europe and to the United States, where LAMCO's 25 per cent partner, the Bethlehem Steel Corporation, receives 2 to 3 million tons per year. In Asia, Japan is an important customer with a growing interest in Liberian ore (Table 6.8).

Table 6.8

*Destination of iron ore shipped from
the port of Buchanan, 1963–6*
(metric tons)

Country of destination	1963 (July–Dec)	1964	1965	1966
Belgium	165,996	826,177	990,460	937,400
France	152,063	359,480	403,200	416,070
Italy	17,611	274,040	963,880	595,460
Japan	—	—	260,430	271,220
Netherlands	—	—	30,820	—
Sweden	—	57,500	146,690	212,040
United Kingdom	—	14,420	352,090	225,810
U.S.A.	653,862	2,660,937	2,397,593	2,746,728
West Germany	797,236	2,682,570	2,771,147	2,702,062
Total	1,786,768	6,875,124	8,316,310	8,106,790

Source: *Liberian Iron Ore Ltd.*, Prospectus (*1966*) *pp. 10–12, and LAMCO J.V., Liberia.*

The average iron content of the mineral shipped from 1963 to 1967 was 67·0 per cent, the average moisture content at the receiving ports decreasing from 7·6 per cent in 1963 to under 6·6 per cent in 1967.

The economic significance of the ports of Greenville and Harper is very restricted in comparison with Monrovia and Buchanan, and the total tonnage handled does not even reach 1 per cent of the national total. The present situation will soon change at Greenville, however, as a result of three factors already mentioned, namely, the immense forest reserves, the expansion of the road system and the construction of new harbour facilities.

The hinterland of Greenville is not only the largest continuous forest area of Liberia but also one of the most sparsely populated regions of the country, so that forest destruction by human influence is almost absent; there are, however, heavy disturbances locally by the large elephant population. Nevertheless, a forest inventory made in the Krahn-Bassa Forest from 1963 to 1967 revealed that on an area of 328,000 hectares (810,000 acres) of exploitable forest there existed a total potential of about 17·25 million metric tons of timber of different characteristics, including export logs, sawn lumber, timber for heavy construction and for core veneer. In the northern zone, a large-scale sawmill with a capacity of 45,000 tons of roundwood per year could be operated for at least fifty years, while in the southern section another sawmill and a veneer or plywood mill with a yearly capacity of 80,000 tons could be run for at least forty years.[1] The production of these mills together with the round logs for export would be shipped through the port of Greenville. Of special importance is the frequency of 'Sikon' (*Tetraberlinia tubmaniana* [J. Leonhard], also called African pine locally). This species is only found in Liberia and is confined to the coastal zone of heavy rainfall. It offers the great and unique advantage of forming large and sometimes even pure stands of trees. On average there are 12 trees per acre (30 per hectare) with a diameter of 16 in. (40 cm.) at breast height and a merchantable volume of 1,057 cu. ft per acre (74 tons [74 cu. m.] per hectare).[2]

Exports of Sikon and other logs via Greenville started immediately after the construction of the Greenville–Zwedru highway in 1967. In February 1968 a total of 1,566 tons was shipped, in March 1,876 tons, in April 1,138 tons and in May 2,235 tons. About 70 per cent of the timber went to France. Monthly exports of 5,000 tons are planned.[3]

The Grebo National Forest in the hinterland of Harper has a suitable exploitation area of 142,000 hectares (350,000 acres), with a merchantable

1 M. *Sachtler and K. Hamer*, Inventory of Krahn-Bassa and Sapo National Forest, *Technical Report No. 7 of the German Forestry Mission to Liberia* (*Monrovia, 1967*) *pp. 2–3.*

2 *Tetraberlinia tubmaniana* (J. Leonhard) "Sikon": a New Timber Species from Liberia, *Report No. 4 of the German Forestry Mission to Liberia* (*Monrovia, 1966*) *p. 6 (mimeographed).*

3 *Information from Greenville port management, June 1968.*

and sub-merchantable volume of about 8,000 tons per sq. km. The minimum potential of the export timber species amounts to 780,000 tons, and in addition there are at least 830,000 tons of core veneer timber available.[1] The future development of Harper port will mainly depend on the exploitation of these timber resources. Unfortunately, geographical problems presented by the state of the road system and technical problems presented by the new port works will restrict the export figures of this commodity for the near future. During the fiscal year 1966–7 log exports from Harper amounted to 4,194 tons, which is less than the tonnage shipped from Greenville within four months of 1968.[2]

Summarising, it may be said that at present the ports of Greenville and Harper have very restricted economic significance, based mainly on rubber and cocoa in the case of Harper and on some palm kernel and a little rubber at Greenville, but that the future prospects based on the great timber resources are promising for both ports.

Robertsport and River Cess, on the other hand, have very limited possibilities of development. This is mainly due to three facts: the lack of protected harbour basins and other port facilities, the absence of commercial farming and mining activities, and the geographical isolation produced by the absence of access roads and of direct road connections with their respective hinterlands.

Development problems

Some of the technical problems of development are closely connected with the geographical characteristics of the Liberian coast. An example is Harper, where an investigation by an American company in 1965 concluded that the port 'was neither well designed nor well constructed, and is at the present time virtually unusable'.[3] The report mentions that in the process of constructing the breakwater, too much rock was excavated from Russwurm Island, 'with the result that during storms, heavy seas wash over the island, causing damage in the warehouse area'. As a consequence, all the traffic is handled by surf boats except for small vessels unloading heavy pieces of machinery or discharging petroleum products. The two warehouses have never been used for the storing of cargo, and

1 Inventory of Grebo National Forest, *Technical Report No. 5 of the German Forestry Mission to Liberia* (*Monrovia, 1967*) *pp. 1–29.*
2 *Republic of Liberia, Department of Commerce and Industry,* Annual Report, 1966–1967 (*Monrovia, 1967*) *p. 70.*
3 *Development and Resources Corporation,* The Development of South-east Liberia (*New York, 1966*) *p. 116.*

timber logs have to be rafted from the harbour basin to the ocean-going ships moored in the open roadstead, so that only 100 to 150 tons per day can be shipped. Because of ground swell, which is considerable at times even within the harbour basin, not even small coastal vessels of the feeder type normally berth alongside but anchor clear of the pier.[1]

In order to solve the technical problems of Harper port, the Liberian Government is undertaking negotiations for the reconstruction of the existing port or construction of a new port in the Harper City area.[2] Apparently, the improvement of the present facilities will not provide an efficient timber outlet owing to the basic size and depth limitations. On the other hand the cost of minimum facilities provided by a new port is estimated at U.S. $5,795,000.[3] A final decision on this question will require a full analysis of the locational, economic and other features of the port and its P.C.D.A. as compared with the estimated construction costs. This analysis should consider the nearness of the new Greenville–Zwedru road, which could be reached by a 100-km. (60-mile) connection from the Grebo Forest, the marginal location of Harper near the eastern border, and the timber potential in the hinterland.

The port of Greenville (Fig. 6.4) like Harper, faces problems which are partly due to disregard of the close interrelationship between technical planning and physical geographic elements, particularly oceanographic features such as wave action, ground-swell action, shoals and other navigational hazards. Heavy wave action caused the design of the break-water to be modified during the construction period in 1956 and extensive maintenance work had to be done at the breakwater and at the head of the pier in 1966–7. At the same time a sandbank at the latter was steadily increasing and needs constant dredging.[4] Wave action not only causes damage to the outside parts of the harbour but is also felt within the basin under strong south-west wind conditions, which fortunately prevail for only about six days a year. The navigational situation is most difficult. 'Ships approaching from the east must sail close inshore, parallel to the coast, to avoid offshore shoal areas, then swing north-east, and finally make a sharp 180-degree turn to enter the harbour and berth in a southerly direction.'[5] In spite of these limitations, the port 'appears to be a well-constructed and well-run facility' so that further modifications will only

1 *Brown Engineers*, Transportation Survey of Liberia (*New York, 1963*) *p. 227*.
2 *Department of Commerce and Industry*, Annual Report, 1966–7, *p. 69*.
3 *Development and Resources Corporation, op. cit., p. 117*.
4 *Department of Commerce and Industry, op. cit.*, Report, *pp. 66–7*.
5 *Development and Resources Corporation, op. cit., p. 114*.

be needed to cope with functional changes caused by the economic development within the P.C.D.A.[1]

The port of Buchanan is only of restricted value in terms of the general economic development of the region as long as the technical facilities for the storing and handling of general cargo and agricultural produce are insufficient. Warehouses, transit sheds, cold storage and handling equipment such as fork-lifts and trucks are indispensable for making full use of the commercial quay built by LAMCO J.V. This problem, however, is juridical rather than technical, in that it was only in 1967 that a National Port Authority was created in Liberia by an amendment of the Public Authorities Law; this still awaits implementation, so that no agreement can be made for the operation of the commercial port. Up to the present, all the Liberian ports with the exception of Greenville are managed by private companies, the Government having little or no jurisdiction.

The principal development problem of Buchanan in its role as one of the world's principal iron-ore shipping ports is the growing size of ore-carriers and other bulk freighters. When the ore-loading section was designed in 1957–9, a maximum deadweight of 46,000 tons was thought adequate and it was expected that the majority of the carriers employed would be in the 29,000–38,000 tons d.w.t. class.[2] In fact, the size has grown constantly and faster than expected, with the result that the average cargo per ship increased from 31,980 tons in 1964 to 34,100 tons in 1965, to 39,740 tons in 1966, and reached 41,720 tons in 1967.[3] The largest ship handled at Buchanan up to 1967, the M.V. *Gloric* of 82,547 tons d.w.t., could only take 69,130 tons of mineral to Philadelphia because of draught limitations at both the ports concerned.[4] This problem will become more and more acute at both Buchanan and Monrovia considering that carriers of 130,000 tons d.w.t. and more are being built and that thirteen ore ports, including seven receiver ports, are deeper than Buchanan, with a present maximum of 18 m. (59 ft) at San Nicolás in Peru.[5]

Problems originating in the inadequacy of facilities at the free port of

---

1 *The port of Greenville was designed for medium-sized banana ships and not for timber exports. See W. Schulze,* Economic Development and the growth of Transportation in Liberia, *Sierra Leone Geographical Association, Occasional Paper No. 1 (Freetown, 1965) p. 8.*

2 *Brown Engineers, op. cit., p. 199.*

3 *Figures calculated from diagrams by LAMCO J.V.*

4 LAMCO News, *published by LAMCO J.V., no. 6/7 (1967) 6.*

5 *Information by Mr O. Goeransson, Port Superintendent, Buchanan, LAMCO J.V., Apr. 1968. Experiments have since been conducted in 'topping-up' large carriers by smaller carriers at sea. Technical studies have also been initiated for a new ore-loading pier.*

Monrovia are less grave than they were in 1961–2, when European and American Conference Lines imposed surcharges on Monrovia cargo of up to 60 and 40 per cent respectively on account of the excessive delays experienced. In some cases such delays lasted from fifteen to thirty-one days, with as many as fifteen to twenty ships waiting at anchor outside the harbour.[1] This situation was mainly due to the fact that from 1951 to May 1962 the Liberia Mining Company occupied some 244 m. (800 ft) of the commercial quay and up to June 1962 there was no special pier for oil tankers. When the L.M.C. finger pier and the oil jetty were completed, freight surcharges were suspended in December 1962. A specific technical problem is the protection of the steel pipes under the wharf apron and the Mesurado Bridge against corrosion by salt-water action. As recommended by the Sly–Hedden Report of 1954, the largest cathodic protection system of Africa was installed in 1957–9 at a total cost of U.S. $494,175.[2] A number of smaller problems remain to be solved if the harbour is to expand as a free port by increasing the number of bonded distribution and manufacturing facilities, and if it is to be prepared for the growing size of bulk carriers and for increased fish production. At present some 20,000 tons of fish are handled per year, including great quantities of tuna fish which are transhipped to other countries via the free port because of the absence of a market in Liberia. If the plans for deepening the channel to the fishing pier to 7 m. (23 ft) and for widening the channel and the turning basin are carried out and additional cold storage is provided, fish production could be doubled within a few years.[3]

In addition, the following improvements would contribute to the expansion, better utilisation and greater efficiency of the free port:

(a) an increase in the size and depth of the dredged area in the harbour, with emphasis on the channels to the iron-ore piers where at least 13·7 m. (45 ft) is required;

(b) an extension of the marginal wharf for another 305 m. (1,000 ft) towards the south breakwater in order to obtain additional berthing space for dry cargo and passenger ships;

(c) movement of the coastal vessels and the latex tankers away from the commercial quay;

(d) acquisition of additional land adjacent to the free port in order to

1 *Brown Engineers, op. cit., p. 188.*
2 *Ibid., p. 155.*
3 *Department of Commerce and Industry,* Annual Reports, 1965–1966 *and* 1966–1967, *and information by port management, Monrovia.*

accommodate warehouses, factories and offices of companies interested in taking advantage of the free-port function of Monrovia harbour; and

(e) provision of additional public warehouse and passenger facilities and widening of the existing road and drainage systems.

Most of the foregoing suggestions are included in an action programme of the Liberian Government of 1966, so that a solution of the technical development problems can be expected in the near future.

The geographical problems of development facing the Liberian ports are less numerous than the technical problems but more difficult to overcome. Their solution requires more scientific research, greater intensity of planning for longer periods of time and higher capital investments. This is not only true for the road construction required in many sections of the country, especially along the coast, but also for the development of new manufacturing industries and additional cash-crop production or the marketing of forest products.

Agricultural projects of concern to port development include the increase of tobacco cultivation near Ganta and on the L.A.C. plantation, and the expansion of coconut and oil-palm plantations. Palm kernels have represented an important export commodity for many years; they were exported from Africa for the first time in 1850 by the Liberian Samuel S. Herring from Buchanan.[1] The scattered distribution of oil palms, their low productivity and the fluctuation of prices (U.S. $79·60 per ton in 1961, $144·68 in 1965)[2] have hindered an increase of production in the past. After 1960, large oil-palm plantation projects with a total area of some 16,000 hectares (40,000 acres) were started at New Cess and in Maryland County. They will result in an increase of the export volume of Buchanan and Harper, in whose P.C.D.A.s only about 30 per cent of the country's palm-kernel production takes place at present.[3]

A special geographical problem in the development of Harper port is the missing road link from Karloke to Tatuke (32 km., or 20 miles). At present, therefore, Harper is still not connected by land route to the rest of Liberia, and until the gap is closed (in the near future) the P.C.D.A. is actually much smaller than that shown in Fig. 6.1.

1 *Sir Harry Johnston*, Liberia (*London, 1906*) 1 *405*.
2 *Liberian Development Corporation and Checci and Co. Contract Group*, Production of Edible Oils in Liberia (*Monrovia, 1967*) *p. 12*.
3 *Robert Cole*, Traditional Agriculture Exports: Sources and Markets, *Northwestern University Economic Survey of Liberia, Staff Paper No. 7* (*Evanston, Ill., 1962*) *map 2*.

In addition to the development problems resulting from the characteristics of the individual ports, there are problems resulting from the physical geographical conditions of the country, of which three examples may be given. The first concerns the frequency of dolerite (diabase) and other solid rock along the coast, including the centre of the Monrovia harbour basin. Removal work started in January 1967 in order to dredge the channel to the iron-ore piers to a depth of 12·7 m. (42 ft). A seismological survey was necessary but work could not be completed until the end of the year because of the physical difficulties.[1] A second example concerns the soft alluvial swamp soils and marshes which represent about 4 per cent of the Liberian territory. Planning of the 1·5 million-ton ore storage yard at Buchanan on reclaimed swampland required extensive field and laboratory investigations and special foundation works. Recordings of pore water pressure gauges installed in the south-west end of the stockpile are still going on.[2] Similar problems occurred near the port of Monrovia when the Bong Mining Company constructed a railway bridge on Bushrod Island. An earthen dam leading to the bridge sank down into the soft marshland so that great quantities of additional material were needed to reach the necessary height.

The third example arises from the high rainfall, which along the north-western coast exceeds 635 cm. (250 in.) a year at times. The rainy season not only restricts the collection of palm fruit but also affects the unloading of rice imports in Monrovia as well as the iron-ore mining and shipping operations. Ore from the stockpile slips down and affects loading and transport equipment if not properly protected. Slippage on the drive drums stops the conveyor belts, and high moisture content reduces the market price of the ore.[3] This is particularly significant with the Mano river ore, which sometimes contains over 10 per cent of moisture. For this reason, in both the ore ports particular attention is paid to keeping the water content as low as possible by proper shaping and drainage of stockpiles, by expeditious dumping of trains upon arrival, and by discontinuing the loading of ships during rain.[4] The close interrelation between iron-ore shipments and monthly rainfall totals is shown in Fig. 6.6.

1 *Department of Commerce and Industry*, Annual Report 1966–1967, pp. *62–3*.
2 O. Goeransson, The Port of Buchanan, *LAMCO Newsletter No. 7 (Stockholm, 1967) p. 11.*
3 *Ibid.*, and *Liberian Iron Ore Ltd* Prospectus *(Stockholm, 1966) p. 31. Iron ore is customarily sold on the analysis of natural iron, which equals the percentage of dry iron × (100 − percentage of moisture).*
4 *Goeransson, op. cit., p. 16.*

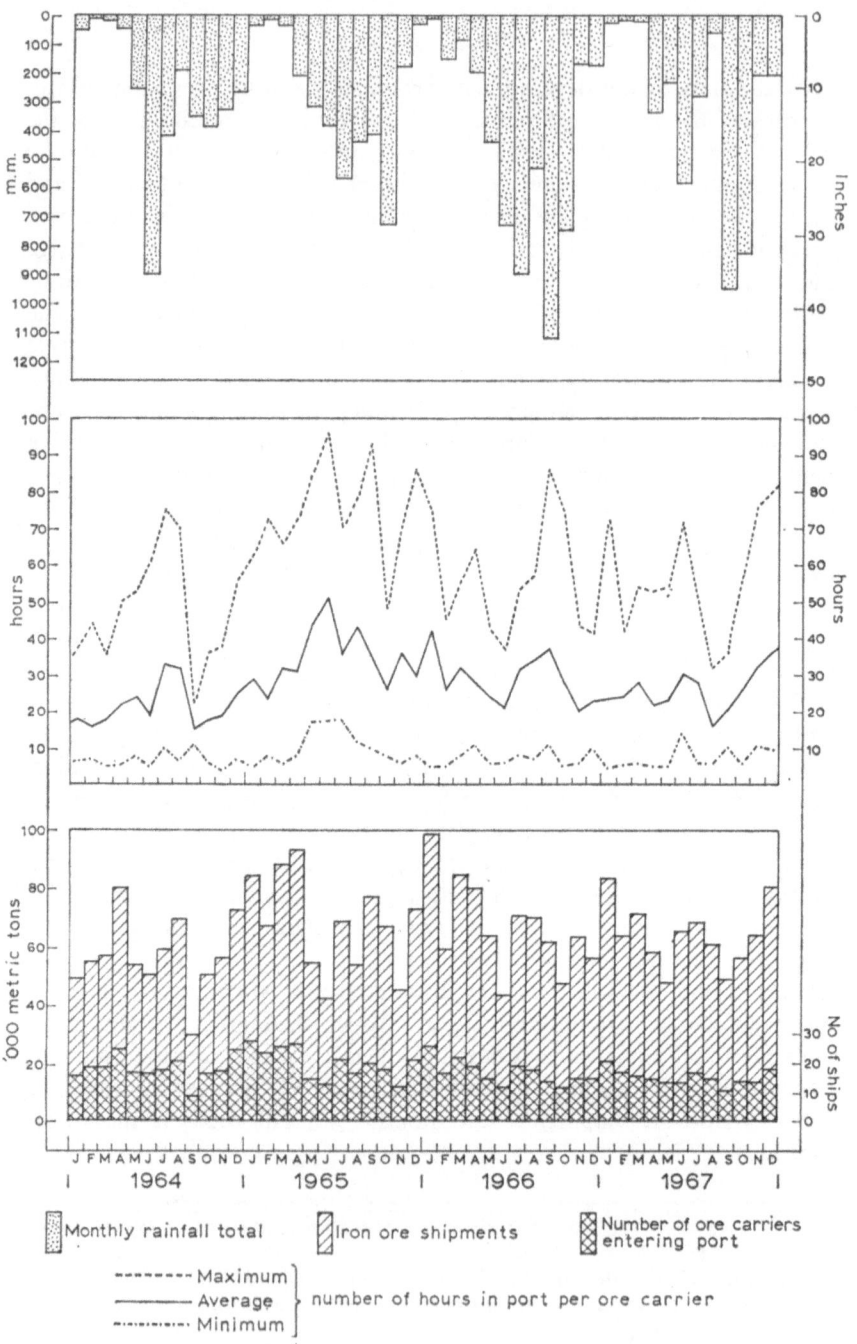

Fig. 6.6   The port of Buchanan: monthly variations of rainfall, iron-ore shipments, and movements of ore-carriers, 1964–7

Summary of development problems

Each of the Liberian ports has its specific problems of development which result from the technical characteristics of the ports concerned or from the geographical characteristics of the individual sites, their hinterlands or the country in general. Some of the problems which affect all the Liberian ports to a greater or lesser degree are the following:

1. Problems resulting from the geographical location of the country close to the Equator. The tropical climate implies in this region a heavy precipitation of up to 35·6 cm. (14 in.) a day during the rainy seasons, high relative humidity thoughout the year and high temperatures, so that special facilities and precautions for the handling of perishable goods and even for iron ore are necessary.

2. Oceanographic problems presented by the many reefs and shoals which make navigation difficult, and silting and ground swell which affect nearly all the harbour basins.

3. Strong corrosion by the warm salt water of the Atlantic (76° to 82°F.) requires expensive protection systems for steel structures or resistant construction materials.

4. Soft alluvial soils near the coast handicap heavy construction works in the port and ore stockpiling areas.

5. Technical problems of general concern are the insufficiency or absence of warehouse and cool storage space, of cranes and handling equipment for general cargo and logs, of passenger facilities and of repair facilities. The Greenville tug *Bell Rock* had to be sent to Abidjan for repair in 1967 because of the absence of a sufficiently large dry dock in the country, with the consequence that it was absent for three months and some U.S. $40,000 had to be spent.

6. The present draught is insufficient in all the harbour basins, including the two ore ports for which the growing size of the ore-carriers is a matter of special concern.

7. Some of the technical problems originate in the fact that the Liberian Government has delegated port operation to independent agencies or private companies which did not have the necessary agreements, initiative, technical knowledge or financial means for the timely expansion or the proper management of the ports. 'The policy has not worked well in all cases (e.g. private control of the Free Port of Monrovia).'[1]

1 R. W. Clower, G. Dalton, M. Harwitz and A. A. Walters, Growth without Development (*Evanston, Ill., 1966*) p. 81.

Table 6.9

Port development in Liberia: a geographical and technical analysis

	Robertsport	Monrovia	Buchanan	Greenville	Harper
Geographical Analysis					
P.C.D.A. of port	1	3	2	2	2
Rainfall	1	1	1	1	2
Relative humidity	1	1	1	1	1
Iron-ore deposits	1	3	3	0	2
Iron-ore mining	0	3	3	0	0
Rubber production	0	3	2	1	2
Palm-kernel production	1	3	2	1	1
Coffee production	1	3	1	1	2
Cocoa production	0	3	1	1	2
Forest reserves	0	3	2	3	2
Fishery	1	3	1	1	2
Manufacturing industries	0	3	1	0	1
Electric energy	1	3	2	1	1
Road mileage	1	3	2	1	1
Road density	2	2	1	1	1
Railway mileage	0	2	2	0	0
Railway density	0	1	2	0	0
Population number	1	3	2	1	1
Population density	2	2	2	1	1
Persons in money economy	1	3	2	1	1
Number of towns	1	3	3	1	1
Index of geographical Development	16	54	38	19	26
Technical Analysis					
Size of harbour basin	0	3	3	2	1
Draught	0	2	3	1	1
Accessibility to ships	0	3	3	1	1
Security in basin	0	3	3	2	1
Navigation aids	0	3	3	2	1
Tugboats	0	3	3	2	0
Commercial quay	0	3	3	2	1
Cranes	0	2	2	2	0
Sheds, warehouses	0	3	0	2	1
Usability of sheds, etc.	0	3	0	3	0
Special facilities:					
Iron-ore handling	0	3	3	0	0
Rubber and latex storing	0	3	0	0	0
Fishing pier	0	2	0	0	0
Tanker pier	0	3	2	0	0
Cold storage	0	3	0	0	0
Index of technical structure	0	42	28	19	7

Key
0 = *non-existent*
1 = *poor*
2 = *fair*
3 = *good*

8. Problems which result from the geographical features of the respective P.C.D.A.s are more varied but with the exception of Monrovia the underdeveloped infrastructure (roads, railways, energy, etc.), the low population density, the low degree of industrialisation and urbanisation, the small quantities of agricultural and lumber production, affect all the ports.

The common features as well as the diversity of the Liberian port development is shown by a geographical and technical analysis in Table 6.9.

As a result of the double analysis in Table 6.9, which adapts a method applied by W. A. Hance,[1] a tentative assessment of the five principal Liberian ports may be made. At a glance it can be recognised that the development problems of Harper are mainly of a technical nature, while the geographical features of its P.C.D.A. are somewhat more favourable than those of Greenville. Robertsport has a very low index of geographical development and has none of the facilities of a modern port. Even if the values assigned to the different factors require some weighting according to their relative significance, it is obvious that the present, and also the future, opportunities for development are relatively more favourable when the addition of the two indices results in a high total. The total index, which might be called the Port Development Index (P.D.I.), is highest for the free port of Monrovia with a value of 96, followed by Buchanan (66), Greenville (38), Harper (33) and Robertsport with only 16. The great difference between the highest and the lowest P.D.I. is a clear expression of the ranking and of the problems existing at the ports. The economic significance in terms of the call/cargo/value quotient mentioned above is not directly related to the P.D.I. and may develop even more in favour of Buchanan when the expected increase of mineral exports and the plans for a West African steel mill located at Buchanan come true. On the whole it is certainly true that 'the most dramatic shock to the Liberian economy during the 1940–1950 decade was the construction of the Free Port of Monrovia . . .'[2] It is probably also true that there are few developing countries where rapid economic growth has depended to such a great extent on the construction of modern ports.

1 *Hance, op. cit., pp. 289–91.*
2 *R. U. McLaughlin,* Foreign Investment and Development in Liberia *(New York, 1966) p. 181.*

7. The Development of the Port of Abidjan and the Economic Growth of Ivory Coast

Mireille Bouthier

Mme Mireille Bouthier is on the staff of the Institut de Géographie Tropicale in the University of Abidjan, Ivory Coast.

The port of Abidjan is a relative newcomer amongst tropical African seaports, since it was only in 1950 that the wharf at Port Bouet gave way to the deep-water port as a result of the cutting of the Vridi Canal which links Ebrié Lagoon with the open sea. In that year the total cargo traffic handled at the four lighterage wharves of Ivory Coast – Port Bouet, Grand Bassam, Sassandra and Tabou (Fig. 7.1) – reached 608,000 tons. In 1951 the traffic of Abidjan alone amounted to 700,000 tons, and ten years later reached 1,700,000 tons. In 1967 Abidjan handled over 4 million tons of goods. This spectacular growth of traffic, largely unforeseen twenty years ago – the most optimistic estimates suggested a figure of 600,000 to 850,000 tons within ten years of the opening of the new port – has been paralleled by the continuous development of port facilities and is closely linked with the rapid expansion of many forms of economic activity, so that the opening of the port has appeared to be the key to the growing prosperity of the country and one of the secrets of the so-called 'economic miracle' of Ivory Coast.

This study attempts to focus attention upon three interrelated aspects of the port of Abidjan and its relationship with the expanding economy of the country which it serves. Firstly, the specific circumstances involved in the initial development of the new port are examined; these include historical and technical factors which help to explain why fifty years elapsed before a deep-water port was established, although during that time external trade developed considerably in spite of the very restrictive limitations imposed by the lighterage wharves. Secondly, the consequences of the opening of the deep-water port upon the various sectors of the Ivory Coast economy are discussed, with special reference to the role of the seaport in particular and of transport facilities in general in the process of economic growth from a stage of underdevelopment to the point of 'take-off'. Thirdly, some attention is given to the structure of the external

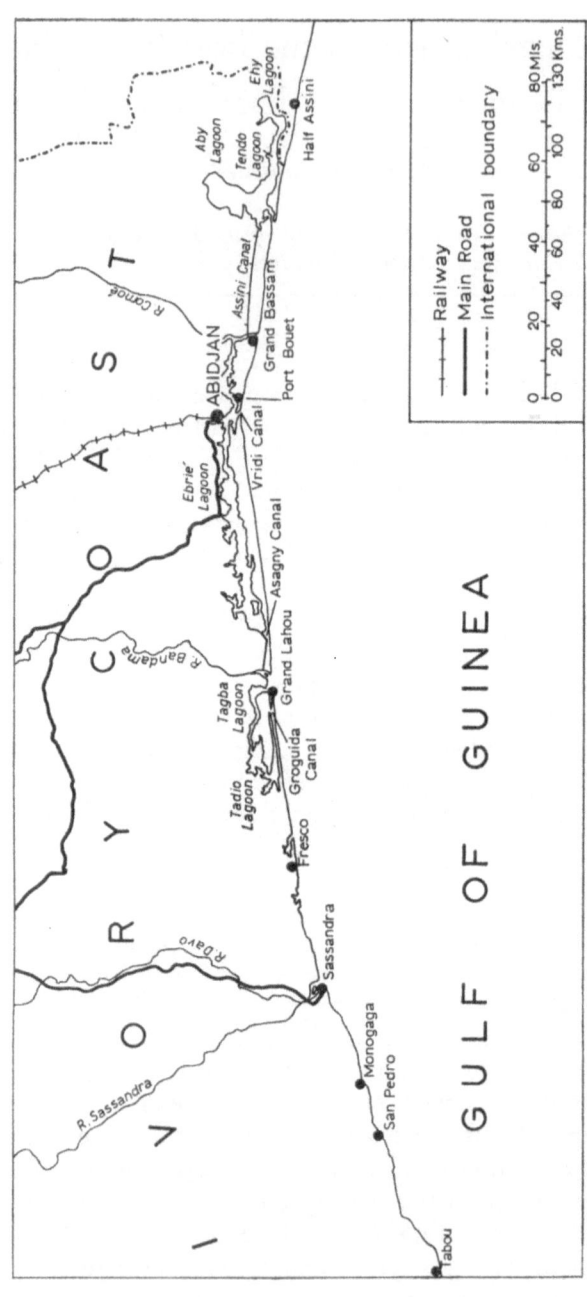

Fig. 7.1 Location of the port of Abidjan

trade of Ivory Coast, which reflects both the rate of current growth and its nature, and to the economic opportunities which this growth may provide in the future.

Origins and development of the port

The coastal environment and the problem of port sites

The selection of the site of the deep-water port of Abidjan arose from historical circumstances and from the particular interest of a few individuals rather than from obvious natural advantages. In fact no mention is made of Abidjan before 1898, although regular commercial transactions had taken place along Ivory Coast (then known as the 'Côte des Dents') since 1863; explorers and traders had recorded points of relatively easy access along the coast since the seventeenth century, but Abidjan was never included. The explanation lies in the morphology of this section of the West African coast. To the west, the only indentations in the rocky and forbidding shoreline are provided by the mouths of the Tabou, San Pedro and Sassandra rivers, the latter providing a splendid harbour sheltered both from sand accumulation and from the offshore surf. The natural advantages of Sassandra harbour might have proved to be a determining factor in the pattern of port development in Ivory Coast, but for the fact that navigators habitually avoided this stretch of coast, known as the 'Côte des Mal Gens', and anchored instead further east off the 'Côte des Bonnes Gens'. This consists of a sandy beach which stretches without major interruption as far as Cape Three Points (Ghana) and is formed by a bar which separates the coastal system of lagoons from the open sea. Access is difficult because of the surf and the strong west–east coastal current. The infrequent breaches in the coastal sand bar were all used between 1700 and 1900 as natural anchorages, although they were dangerous in navigational terms: they included Assini, at the mouth of Aby Lagoon; Grand Bassam, at the mouth of the Comoé river east of Ebrié Lagoon; and Grand Lahou, near the mouth of the Bandama river. During this period Abidjan was only a small fishing village on the Ebrié Lagoon, with no direct communication with the open sea and therefore of no interest in terms of external trade; but the settlement was well placed to benefit from the growth of trade and economic development throughout the lagoon zone. The first coffee, cocoa and oil-palm plantations were established here, and early timber exploitation was favoured by the possibility of floating logs to the exporting ports; and it was from this zone also that the first surveys of the interior set out.

Around the turn of the century the explorations of Treich-Laplène and

Binger were paralleled by the commercial enterprise of traders such as Verdier, upon whose initiative the port of Grand Bassam was chosen as the seat of the Resident of the colony of Ivory Coast in 1893. Trade was concentrated at this point, therefore, in spite of the rather awkward navigational conditions involved. A few years later, in 1898, when the French Government decided to build a railway linking the interior parts of Ivory Coast with the sea, the Houdaille mission suggested Abidjan as a suitable site for the seaward terminus of the railway and for the development of an ocean port.[1] Many factors affected this recommendation, amongst which the influence of an important trading company established in the locality – the Compagnie d'Afrique Occidentale – was doubtless of considerable importance.

Advantages and technical problems of the site of Abidjan
The decisive factors involved in the selection of Abidjan as the site of a new port in preference to a variety of other possibilities were clearly physical and technical considerations. Ebrié Lagoon (Fig. 7.2) provides at this point a splendid stretch of sheltered water with depths exceeding 10 m. (33 ft) over an area of 800 hectares (3·1 sq. miles) – Dakar harbour has only 50 hectares (0·2 sq. miles) at this depth – whilst other parts of the lagoon are comparatively shallow. To link the lagoon with the sea, the obvious alignment for a canal through the coastal sand bar lay in the vicinity of the village of Port Bouet, where the bar is only 800 m.–1 km. (0·5–0·625 mile) wide and presents no serious technical difficulties. Another important feature was the presence of a submarine trough, known since the seventeenth century, some 400–500 m. (440–550 ft) deep a few hundred metres offshore from Port Bouet. The possibility was considered of using this unusual feature, known locally as the 'Trou sans fond', as a trap for sand moving along the coast. Under the influence of the swell, which attacks this stretch of the West African coast at an angle of 15–25 degrees, some 1·5 million cu. m. of sand is moved along the shore from west to east each year. This inevitably obstructs all coastal inlets, natural or artificial, but engineers thought that a canal through the coastal sand bar might be constructed in such a way that sand would be diverted into the 'Trou sans fond' and the canal would be kept clear.

In spite of these apparently favourable circumstances, the development of a deep-water port at Abidjan was delayed for half a century. The building

1 *For further historical material and details of the Houdaille mission, see G. Rougerie, 'Le problème des débouchés maritimes de la Côte d'Ivoire', Bulletin de l'I.F.A.N., vol. 12, no. 3 (1950).*

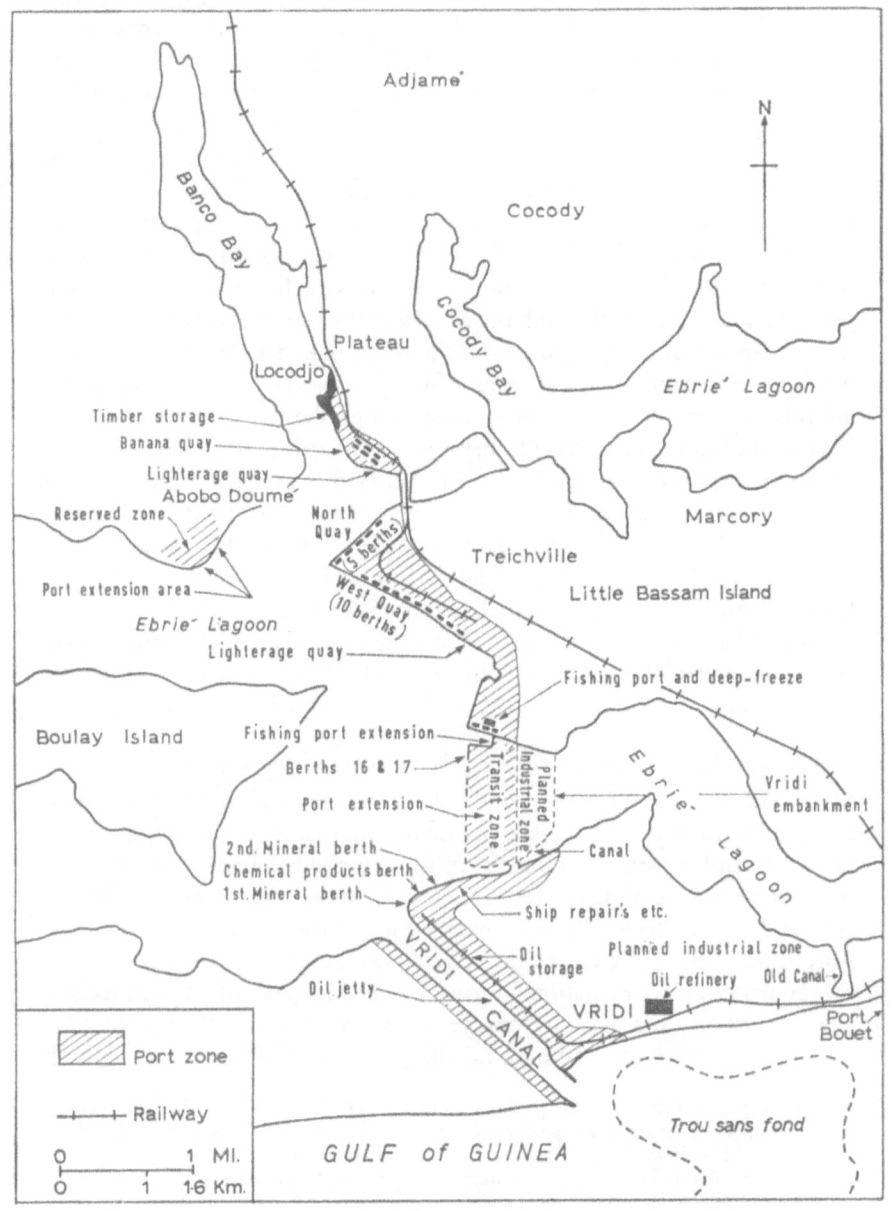

Fig. 7.2   The port of Abidjan: layout and facilities

of the railway and the excavation of the canal began simultaneously in 1904, but whereas the railway made steady progress the port works were a complete failure. By 1907 it was realised that the 'Trou sans fond' was incapable of preventing the rapid silting-up of the canal, and that the technical aspects of the problem were far more complex than had been anticipated. Further studies were undertaken, continuing until 1936,[1] with special reference to the direction and strength of the coastal currents; in order to encourage the natural removal of any sand deposited at the entrance, a canal 370 m. (1,214 ft) wide was recommended, narrowed at its seaward end by two jetties the presence of which would accelerate the seaward flow of water through the canal and thus create a natural dredging device capable of carrying the sand for more than a mile out to sea. The north-west to south-east alignment of the proposed canal was designed to aid this process, the seaward end being located slightly to the west of the 'Trou sans fond' so that sand transported eastwards by the coastal current would be deposited therein. In order to verify the results of this theoretical scheme, a model incorporating the precise conditions of the coastal sand bar, the submarine trough, the sand movement, the swell and the currents was developed in a hydrological laboratory at Delft. The reasons for the failure of the Port Bouet channel in 1907 were amply demonstrated, and the scientific precision of the new project was confirmed. Work began on the canal project shortly before the outbreak of the Second World War, and the completed channel, known as the Vridi Canal, was opened in 1950.

Lighterage wharves and their effect on the economy of Ivory Coast

In the absence of modern port facilities traders and shipowners were obliged for fifty years to utilise the inadequate facilities of the lighterage ports.[2] Grand Bassam operated from 1901 in conditions so unsatisfactory that the maritime archives form a continuous record of complaint. A record total of 169,000 tons of cargo was handled at the port in 1929. Port Bouet, situated close to the railway bridge across Ebrié Lagoon and somewhat safer in navigational terms, reached its peak of 253,000 tons in 1948. Inevitably the high percentage of damaged goods, the loss of time incurred by vessels and the high handling costs were reflected in the high

1 *On the technical problems of the construction of the canal, see R. Pellenard-Considère, 'Naissance du port et de la ville d'Abidjan', Cahiers Charles de Foucauld, vol. 36 (1954).*
2 *After the failure of 1907, several alternative port sites were suggested, e.g. Grand Bassam, Sassandra and the mouth of the Comoé river. The most serious technical study was undertaken by the engineer H. Noël, in 1919, who supported the Abidjan project.*

cost of goods, a factor undoubtedly prejudicial to the economic development of the country.

In 1923 the Chamber of Commerce estimated that the traffic and the population of Ivory Coast no longer justified the creation of a port similar to that at Lagos, which was moreover then considered positively harmful to the Nigerian economy. This argument seems to have represented much more an attempt to justify the failure of the 1907 project and the disappointment caused thereby than an objective evaluation of the economic potentialities of the country. If the lack of export facilities did not totally impede the economic development of Ivory Coast in the 1920s, it held it appreciably in check. The Chamber of Commerce was obliged to admit that in 1925 an important project for the development of groundnut cultivation in the north of the country had had to be postponed because of the congestion at Grand Bassam. This underlined heavily the basic problem of a country with a very considerable economic potential but with transport facilities so inadequate as to constitute a stranglehold on production and movement.

Nevertheless, export figures continued to rise in the 1920s and 1930s, and again after the war; growth reflected mainly increasing output of plantation products such as cocoa (22,000 tons in 1930, 43,000 in 1939), coffee (unknown before 1930, 28,000 tons in 1939, 55,000 in 1948) and bananas (14,000 tons in 1939). Timber exports increased likewise (91,000 tons in 1930), then declined somewhat but regained their predominant place in 1948. Exports thus showed a reasonable diversity in 1948, and imports also continued to grow. The stage of the colonial trade economy was passed in 1900, and thereafter imports of alcohol and textiles were of less significance than tools, machinery and building materials, oil products and rice; furthermore the role of France as a source of supply became preponderant. There is evidence, however, that in the 1940s trade was tending to level out because the lighterage wharves were working to capacity and the opening of the new port in 1950 occurred just in time to prevent complete commercial asphyxia.

Facilities and equipment[1]

When the Vridi Canal was opened in July 1950, vessels entering the port anchored in the open waters of Ebrié Lagoon, since no permanent port

1 *For general documentary and statistical material, see* Port d'Abidjan: rapport annuel (*Direction du Port*); *and* J. L. Tournier, '*The Port of Abidjan*', Dock and Harbour Authority, *vol. 38, no. 449 (1958).*

facilities had been developed; a temporary lighterage quay was provided, and the expansion of facilities on a large scale was begun. Finance was obtained from a variety of sources: France provided the cost of the Vridi Canal (6,000 million francs), whilst investments in the port, totalling 19,000 million francs, were divided between France (12,000 million) the Ivory Coast Government (5,000 million) and the port authorities (2,000 million).

The Vridi Canal only permits the entry of ships of moderate size (30,000 tons) since its depth reaches 15 m. (50 ft) in the centre but does not exceed 10 m. (33 ft) along the margins. On the landward side of the canal the various parts of the port are located around the shores of Ebrié Lagoon. Banco Bay, west of the residential peninsula of central Abidjan, is the location of specialised facilities for the storage and export of timber (Plate 5): the Locodjo timber storage area (65,000 sq. m. [699,400 sq. ft], concrete-paved) is flanked by a 380-m. (405-yd) lighterage quay equipped with two 20-ton cranes. Sixteen vessels, anchored in the bay, can be worked simultaneously. South of the timber zone is a banana quay, with specialised storage facilities, where two vessels can be worked. The heart of the port is formed by the 2,000 m. (2,190 yd) of deep-water quays which form a right-angle on the western side of Little Bassam Island and which are well equipped with cargo-handling, storage and transport facilities. Thirteen vessels can be worked simultaneously. The North Quay offers five berths and five transit sheds with a total area of 28,000 sq. m. (301,300 sq. ft); the easternmost berth is specially equipped to handle grains unloaded for Abidjan flour mills, and the westernmost is the passenger terminal. There are ten berths on the West Quay (Plate 6), beyond which is a lighterage quay serving local traffic only. Immediately to the south lies the fishing port, with a 400-m. (438-yd) quay opened in 1962; facilities here have already reached saturation point and are scheduled for considerable extension. On the eastern side of the lagoon entrance to the Vridi Canal are located a mineral berth designed to handle 800,000 tons of manganese ore a year, and an oil berth equipped to service tankers drawing less than 9 m. (30 ft) of water. Larger tankers, which cannot negotiate the canal, are worked by pipeline whilst anchored offshore.

To meet increases in traffic a programme of port works was begun in 1965 designed to infill that part of Ebrié Lagoon between the south-western tip of Little Bassam Island and Vridi. This scheme will permit the development of ten new deep-water berths backed by a 40-hectare (0·1 sq. mile) transit zone, and, by linking the Vridi area directly with the core zone of the port, will allow the vast area lying between the Vridi

Canal and the 1907 canal to be utilised for industrial development. Thus the efficiency of the port is likely to maintain its upward trend. Whereas in 1964 the proportion of vessels entering the port without waiting was 61 per cent, by 1966 this figure had risen to 75 per cent despite a 19 per cent increase in traffic. Costs per ton handled have declined, the business handled by the port increases, and the possibilities of reinvestment continue to grow.

Economic effects of the new port

The modernisation of the city of Abidjan

The development of a new port invariably produces widespread effects – economic, social and perhaps psychological.[1] In the case of Abidjan, private capital and energy has been mobilised which, together with the investment of public funds, has opened the way towards a more effective development of the country and towards a higher level of living for the bulk of the population. Perhaps the most immediate and most spectacular effect, however, has been the transformation of Abidjan itself from a small 'colonial-style' town into a modern capital city. Increased circulation of people and commodities between the port and industrial areas of Little Bassam Island and the central business and residential areas of the Plateau has necessitated the construction of two concrete bridges (to replace a floating bridge dating from 1929) and of a dual carriageway serving many parts of the built-up area. Industrial firms, banks, etc., have been attracted to the city and have developed large office blocks. Large numbers of immigrants from rural areas and from neighbouring countries[2] have come in search of work and a better life. All these factors have brought about a mushroom growth at Abidjan, which had only 48,000 inhabitants in 1948 and now has 370,000, and also a change in the social and ethnic structure of the town. Foreigners – mainly from Upper Volta and Mali – constitute about one-third of the urban population and as much as 50 per cent of the male urban population aged between fifteen and sixty, and thus play a vital role in the labour situation. Disparity between population increase and expansion of employment opportunities inevitably presents the possibility of the growth of vast areas of shanty towns, inadequately served by transport and other urban facilities, which contrast markedly in physical and social terms with the splendid central-area developments. In fact such problems have not materialised seriously at Abidjan; the growth of the

1 *Samir Amin*, Le développement du capitalisme en Côte d'Ivoire (*Paris, 1967*).
2 *E. Bernus*, '*Notes sur l'agglomération d'Abidjan et sa population*', Bulletin de l'I.F.A.N., *vols 1–2 (1962)*.

town has been reasonably well ordered as a result of an active programme of low-income housing development, the establishment of modern market and social facilities in non-central zones, the provision of water supplies, garbage disposal services, and the extension of the public lighting system. Thus the town has virtually been completely redeveloped during the past fifteen years, and, as a result of the modern utilisation of its unique site, has become a major West African growth point and a symbol of the modernisation of tropical Africa as a whole.

The centralisation of transport media[1]

The opening of the Vridi Canal and the deep-water port was part of the FIDES plan (Fond d'Investissement et de Développement Économique et Social) mounted by the French Government after the Second World War. The object was to equip and develop French Overseas Territories, amongst which Ivory Coast was well placed because of its rich economic potential. The first two plans (1948–56) gave priority to the development of a basic economic infrastructure with special reference to transport. Thus the network of classified roads grew from 3,670 km. (2,295 miles) in 1947 (with 50 km. [31 miles] tarred) to 10,570 km. (6,610 miles) in 1958 (with 600 km. [375 miles] tarred). The two major arteries emerged as the Abidjan–Gagnoa 'coffee road' and the Abidjan–Abengourou 'cocoa road' (Fig. 7.3). The new port permitted the import of heavy machinery and materials for road construction, and thereby stimulated the expansion of the hinterland. The first FIDES plan provided for the improvement of the Sassandra wharf and of the Sassandra–Gagnoa road, on the assumption that most of the western part of the country would continue to lie within the hinterland of the smaller port. In spite of the improvement of facilities, however, traffic at Sassandra has continued to decrease, which underlines further the effect of the creation of a new deep-water port after a half-century of dependence upon inconvenient and expensive wharves. The possession of an effective economic instrument available throughout the year encouraged exporters to make use of it; the hinterland of the port thus rapidly came to include almost the entire country as a result of the extension of the road network.

It would be difficult to overestimate the role of the railway – extensively modernised since 1946 – in the development of the port; subsequently, the extension of the line from Bobo-Dioulasso to Ouagadougou (1955)

1 J. Tricart, Étude géographique des problèmes de transports en Côte d'Ivoire (*Paris, 1963*); S. *Kobe*, Problèmes des transports en Afrique de l'Ouest (*Paris: O.E.C.D., 1967*).

meant that Abidjan became more effectively the port of Upper Volta. Traffic to and from Upper Volta is relatively slight, however, and consists chiefly of imports of salt, flour, food and drink, oil products, metallurgical and construction materials, and of Ivory Coast products such as kola nuts and timber; exports comprise chiefly dried fish, livestock on the hoof,

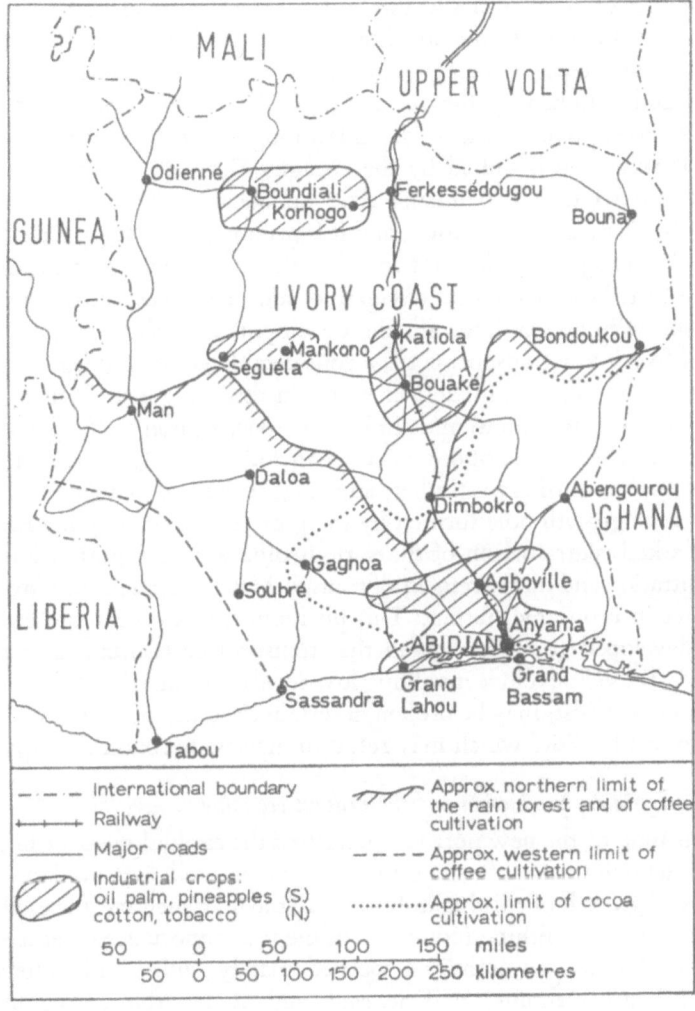

Fig. 7.3 Ivory Coast: patterns of export crop production and surface transport

E

groundnuts and shea-nut butter. The importance of the railway in the transport of Ivory Coast exports has decreased considerably, owing in part to its slowness and lack of adaptation to peak-season requirements. Efforts made in the 1950s to develop road–rail interlinkages met with little success; road competition is very heavy and since 1963 the railway has been running at a loss.

The port of Abidjan lies at the heart of the system of coastal lagoons which extends for almost 300 km. (190 miles) between the Ghanaian frontier and Fresco. Aby Lagoon, in the east, penetrates 35 km. (22 miles) inland and is linked by the 30-km. (19-mile) Assini Canal (1957) with Ebrié Lagoon; over 115 km. (72 miles) long, Ebrié Lagoon receives the Comoé river and is linked by the Assagny Canal to the Grand Lahou Canal into which the Bandama river debouches. The Groguida Canal (1960) further assists circulation, and it is proposed to develop a canal link with Fresco Lagoon, 19 km. (12 miles) to the west. This lagoon system is a splendid method of internal navigation, and it has been completed and modernised in association with the development of the new port. As a result of very low transport costs, a considerable volume of lagoon traffic focuses upon Abidjan; notable elements in this trade include manganese ores (from Grand Lahou) and timber. In 1967, lagoon traffic handled at Abidjan reached a total of 665,454 tons, including 476,235 tons of floating logs, 13,028 tons of sawn timber, and 126,180 tons of manganese.

Thus the growth pole formed by the port and capital city has resulted in a marked centralisation of the various modes of transport, whilst this concentration and improvement of transport facilities has in turn involved increases in economic output. This phenomenon is observable in other underdeveloped countries; given that transport costs decrease as traffic expands, the vicious circle involving low-level economic development and high transport costs may be broken in certain circumstances by investment in transport facilities which may result in increased economic output.

Effects on primary production and the peasant economy

The opening of the new port has permitted the establishment of hitherto unrepresented forms of economic activity as well as stimulating the continued growth of existing industries. In the first category, the establishment of a modern fishing industry is of the first importance.[1] Before 1952 ocean fishing was an activity associated mainly with immigrants from Ghana, Togo, Dahomey and Senegal, and in the Abidjan region was based upon Port Bouet (with sixty large canoes) and Bassam (with eighty

1 M. Lassarat, La pêche en Côte d'Ivoire (*Abidjan, 1967*).

canoes). Lagoon fishing is traditionally practised along the entire length of the system, using 300 large canoes and 1,500 small ones. The annual output of 15,000 tons is traditionally bought by Ghanaian women who dry or smoke the fish and then sell it in Ivory Coast and Ghana. This traditional sector of the fishing industry is at present stationary, but employs some 20,000 persons together with a further 15,000 in fish processing. The modern fishing industry dates from the opening of the Vridi Canal, and is now more important than the traditional sector in terms of output, although only 7,800 persons are employed. A small fleet of trawlers has grown rapidly since the development of fishing and refrigeration facilities (1962) in the new port; in 1955, 14 vessels brought in 5,000 tons, and in 1967, 70 vessels landed 48,000 tons. Abidjan thus emerges as a significant West African fishing port; in 1967, in addition to the locally based operations, a further 20,000 tons were handled by the port (9,000 tons landed by French vessels and 8,500 tons by Japanese) of which some 2,000 tons were canned locally. Tunny-fishing prospects are good, with the opening in 1969 of a refrigerated warehouse with a storage capacity of 3,000 tons at $-20°$ C.

Timber exploitation[1] has been important in Ivory Coast since 1885, but since 1947 the annual average increase in output has been 19 per cent. Growth has been due in part to the facilities provided by the new port, and has been characterised by a progressive westward shift in the location of the major areas of exploitation. The eastern and central regions now produce only 400,000 cu. m. (14 million cu. ft) each per annum, and the north-western region (lying between Gagnoa, Daloa, Man and Soubré) is now the main producing area with an annual output of 1·3 million cu. m. (45 million cu. ft). Timber exploitation in the south-western corner of Ivory Coast is now making rapid advances, and production stands at around 500,000 cu. m. (17·5 million cu. ft) a year; but the ever-increasing cost of transport to Abidjan presents serious problems to which the long-term solution appears to be the opening of a deep-water timber port on the south-west coast. More widely, prospects for the timber industry turn on the question of possible resource exhaustion and the development of reafforestation schemes; important steps in this direction include the establishment in 1965 of an official 'Code Forestier' regulating exploitation permits and requiring the rapid replanting of worked areas by a state organisation (SODEFOR) which began operations in 1968.

Plantation agriculture has expanded by an average of 9·4 per cent per

1 '*Les activités d'exploitation forestière et de transformation du bois en Côte d'Ivoire*', *Report of the Syndicat des Producteurs Forestiers de Côte d'Ivoire, May–June 1967.*

annum between 1951 and 1967, compared with an average of 7·3 per cent for the agricultural sector as a whole. The coffee and cocoa economies are concentrated in the forest zone of the south, and whereas in 1950 only 50 per cent of the population was involved in the cultivation of these specialised crops, the percentage is now much nearer 80. The eastern region, the first to be developed, is today the most important, producing 38 per cent of the total coffee output and 62 per cent of the cocoa. The west-central region as far north-west as Man now takes second place, with 46 per cent of the coffee and 18 per cent of the cocoa output. This emphasis on western and central areas stems partly from good road links with Abidjan, which has also encouraged the cultivation of coffee and cocoa in the savanna country around Bouaké where, in a climatically unsuitable zone, the traditional subsistence economy of the Baoulé people has gradually incorporated elements from the cash sector during the past fifteen years.[1] Apart from coffee and cocoa a variety of other cash crops has been introduced into the forest zone. Peasant growers, in association with state organisations and research groups, are important in terms of pineapple cultivation (output increased 52 per cent between 1966 and 1967), oil-palm plantations (30,000 hectares [115 sq. miles] planted in 1966, including 6,000 village plantations), and coconut plantations which are currently being developed. Rubber plantations are under both government and private control (12,000 hectares [46·3 sq. miles] planted in 1966). In the north, cotton cultivation is being expanded.

An important socio-demographic phenomenon linked with this accelerated economic development of the south is the immigration of large numbers of people from Upper Volta, Mali and Guinea. In 1950 foreigners comprised 2·5 per cent of the rural population and were mainly plantation workers. Fifteen years later they represented 22 per cent of the rural population and 35 per cent of the adult male population. In the plantation areas the proportion of foreigners varies between one-half and two-thirds of the numbers employed. In the east they have largely replaced the local labour force and constitute a distinct class of agricultural wage-earners employed by the planters, who are often former traditional chiefs. In the west, amongst the peoples of the forest and lagoon zones, where the social hierarchy is less distinctive, immigrants have developed plantations and villages in their own right. Whilst the immigrant labour force has thus assisted the rapid economic advance in the south, it has also

---

1 *Amin, op cit., p. 98, suggests that in this region the maximum worked area per cultivator has already been reached; at present this is about 2 hectares (4·9 acres) of plantation cultivation per farmer using extensive methods.*

accentuated the cleavage between the north, where a traditional sub-sistence economy predominates, and the south, where the cash economy is much better developed. The contrast in levels of living between the northern and southern parts of Ivory Coast continues to increase; in the savanna zone, annual income per head increased from 1,400 to 5,200 francs between 1950 and 1965, whereas in the forest zone the respective figures are 11,700 and 25,000 francs. On average, therefore, a member of the Agni or Baoulé tribes earns five times as much as a Senoufo, whilst the income of some 20,000 middle-class planters range between 300,000 and 600,000 francs a year.[1]

The growth of processing industries[2]

It is generally accepted that a port brings into being a variety of transport-orientated industries; this has certainly been the case at Abidjan, which in the course of a few years has become both an important labour pool and a major centre of consumption. In 1950 no general electricity service was available and the few existing industries provided their own power; the restricted range included two small pineapple-canning factories, several oil and soap works supplying local needs, two breweries, several sawmills, a small cotton-spinning mill at Bouaké and an industrial gas plant. Rapid industrial growth has subsequently been assisted by transport development, by the provision of power from the hydro-electric barrage at Ayamé (1959), by the construction of an oil refinery in the port (1965) and finally by the official encouragement of private investment since 1959 by tax relief on priority developments and guaranteed long-term stability of the taxation system. Industrial growth took place at an average annual rate of 15 per cent in the 1950s, but since 1960 this has increased to 25 per cent. These figures relate only to processing industries, most of which are located in Abidjan. Three major industrial groups have emerged. Firstly, light industry concerned with the initial processing of primary products for export: these now include two enlarged pineapple-canning factories, which treated 73,000 tons of fruit in 1967 as against 8,000 tons in 1956; a tunny cannery and a frozen tunny factory; an instant-coffee factory (5,000 tons a year); a factory producing cocoa powder and butter using 25,000 tons of cocoa beans a year; a palm-oil and copra factory which is expanding as the plantations develop; two latex factories; and timber

1 *Amin, op. cit., p. 89.*
2 *For the pre-1960 period, see the following studies published by the Ministère du Plan (Service de la Statistique):* Inventaire économique de la Côte d'Ivoire, 1947–56, *and* Inventaire économique et social de la Côte d'Ivoire, 1947–58.

industries including two sawmills (output 271,000 cu. m. [8·5 million cu. ft] in 1966) and a plywood factory (12,500 cu. m. [447,250 cu. ft]). Secondly, another group of enterprises is involved in the treatment of local produce destined for local consumption: sardines, coffee, rice, soap, matches, cotton textiles and tobacco. The third group comprises light industry designed to serve the local market but based upon imported raw materials: these include flour mills, breweries and other food and drink industries; motor vehicle and cycle assembly, and the production of wire, rails and household goods from imported metals; electrical industries; chemical works; and construction materials.

This pattern of industrialisation poses a number of serious problems in the context of the economic growth of the country as a whole. The concentration of 90 per cent of all industrial development in Ivory Coast in the Abidjan region has created a highly eccentric growth pole which offers few direct benefits to other parts of the country; Bouaké is the only secondary industrial centre. In general terms, the coastal industries may in due course bring about the economic take-off of the interior by progressive penetration;[1] but light industries which concentrate their attention upon an expanding urban market reinforce the sharp contrasts between urban and rural levels of living and ways of life. The lack of basic industry (which would be uneconomic in the narrow confines of the national market) is another major problem, together with that of overdependence upon agriculture; this involves the risk of an early levelling-out of production, whilst increasing imports of raw materials and semi-manufactured goods would tend to upset the balance of payments. Foreign capital presents other problems, in the sense that its dominant position prevents the growth of a truly independent economy; incomes are re-exported, and local élite do not always participate in the administration of business affairs.[2] Nevertheless, certain positive aspects have emerged: imports, although strongly criticised, nevertheless stimulate development by creating demand, and open the way to production in Ivory Coast itself;[3] import-substitute industries are being developed, and include cellulose and paper manufacture and tyre and rubber plants. The

1 A. Bouc, 'Fondements théoriques d'une politique d'industrialisation en Africque de l'Ouest', *in* Problèmes de développement de l'Afrique de l'Ouest (*Dakar: Institut de Science Économique Appliqué, 1966*).

2 Amin, *op. cit., points out that during the 1950–65 period returns on foreign capital invested in Ivory Coast were higher than those on local finance; that external control on reinvestment was absolute; that 46 per cent of the active urban labour force was employed in foreign enterprises; and that 83 per cent of the added value in the productive urban economy was derived from the foreign sector.*

3 A. Hirschman, Stratégie du développement économique (*Paris, 1964*).

port of Abidjan has thus emerged as a growth point, attracting both foreign labour and capital, stimulating work and profit incentives, and developing a wide range of new economic activities. The overall pattern is controlled by firm government planning policies. The growth which has resulted during the past two decades is clearly reflected in the pattern of overseas trade handled by the port.

The pattern of port traffic[1]

The overall pattern of port traffic at Abidjan has been characterised since 1950 by remarkable growth. Taking 100 as the index of traffic for 1951, the 200 index level was reached in 1959, 300 in 1961, 400 in 1962, and 500 in 1965; the 1967 index was 531. Thus, after a relatively slow start, the 1958–66 period showed very rapid growth, whilst 1967 represents a reduction in the rate of increase. The average annual rate of increase during the period 1951–67 was 12·7 per cent, with 1960 (26 per cent) and 1961 (35 per cent) as the peak years.

Volume of traffic handled

Imports and exports are reasonably well balanced at Abidjan (Fig. 7.4). Prior to 1957 imports exceeded exports, as is normal in a developing country in the process of establishing an essential economic infrastructure. After 1957 the situation was reversed: imports continued to increase (1957, 580,000 tons; 1967, 1,600,000 tons) but exports grew even more rapidly (1957, 534,000 tons; 1967, 2,500,000 tons). Imports consist chiefly of cement, oil products and foodstuffs.

Cement took the lead during the most rapid phase of port and urban development, and increased fivefold between 1948 and 1951; in 1967 cement imports totalled 355,000 tons. Imports of crude oil and oil products have increased greatly during the period under consideration (1951, 30,000 tons; 1967, 620,000 tons); since 1954 oil products have emerged as the dominant import commodity, and in 1965 crude oil largely replaced refined products with the opening of a local refinery. Imports of foodstuffs have also increased rapidly and have become more diversified as the average level of living has been raised. Government efforts towards national self-sufficiency in foodstuffs are, however, beginning to show

---

1 *Statistical sources used in this section include:* Bulletin mensuel de statistiques (*Abidjan, Service de la Statistique); Rapports annuels des services de la statistique (Abidjan, 1960–5); 'Analyse du commerce extérieur de la Côte d'Ivoire en 1964, 1965 et 1966'*, Bulletin Mensuel de la Chambre d'Industrie de la Côte d'Ivoire.

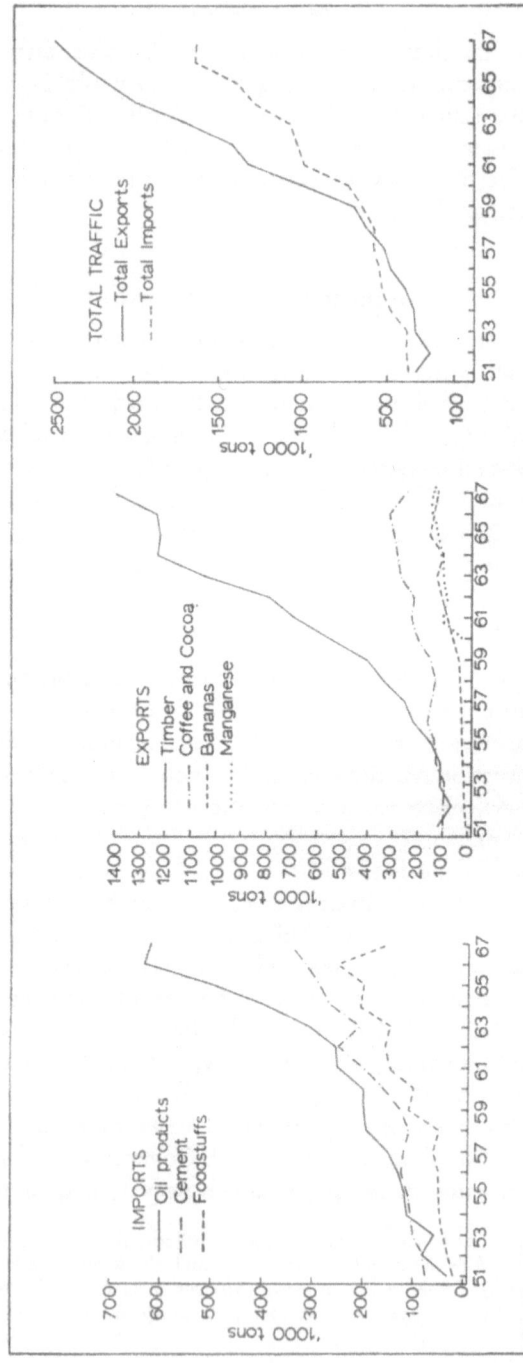

Fig 7.4. *Growth of port traffic at Abidjan, 1951–67*

results: food imports fell by 100,000 tons in 1967, coincident with the first major commercial rice harvest.

Unprocessed timber now forms 60 per cent of the volume of exports, and tonnages have doubled every three years since 1954 (1967, 1,400,000 tons). Timber exports are thus primarily responsible for the rapid growth of export tonnages, and have necessitated constant modification of port facilities to reduce congestion and improve shipping turnround times. Three other export commodities are of fundamental importance at Abidjan: coffee, cocoa and bananas. Coffee and cocoa exports, taken together, have increased threefold from 113,000 tons in 1951 to over 300,000 tons in 1966, whilst banana exports have increased during the same period from 16,000 to 140,000 tons. Exports of palm kernels and palm oil have remained relatively stationary, and rubber and cotton shipments reached only 5,500 and 9,500 tons respectively in 1966; exports of pineapples and their products (canned fruit and juice) have risen from 1,100 tons in 1955 to 48,000 tons in 1967. Since 1960 manganese exports have become important.

Value of traffic handled

Prior to 1958 exports consisted primarily of coffee and cocoa, which represented 85 per cent of the value of goods shipped overseas (Table 7.1). This colonial-type pattern resulted partly from the highly favourable world market situation for these commodities until 1960, and also from the fact that France provided a ready market for the greater part of the annual output. After 1958 exports have become more diversified, and coffee and cocoa now represent approximately 50 per cent of the total value of exports. Timber represents 25 per cent, bananas and pineapples 6 per cent, and manganese and diamonds 2 per cent.

There is thus evidence of diversification of exports in terms of primary products and certain processed goods (pineapples, cocoa, timber), but the overall economic structure remains essentially that of an underdeveloped country, since 65 per cent of the value of external trade is provided by plantations, and timber products still provide almost three-quarters of the total volume exported. Exports of manufactured goods remain insignificant, and the broad economic structure thus remains in qualitative terms little different from what it was in 1951. Furthermore, deterioration in the world market situation in respect of coffee and cocoa, and to a lesser extent in relation to bananas and timber products, is a serious problem affecting the economic prospects of Ivory Coast and the future traffic of Abidjan.

Table 7.1

Value of principal commodities exported through the port of Abidjan, 1951–66, by percentages of total exports

Year	Coffee	Cocoa	Timber	Bananas	Pineapples*	Palm kernels	Minerals†	Others
1951	52·2	35·2	6·1	1·5	0·2	2·2	—	2·7
1952	56·4	32·8	3·3	1·7	0·4	0·9	—	4·5
1953	46·9	41·3	4·8	1·9	0·4	1·8	—	2·9
1954	60·4	32·3	3·7	1·2	0·2	0·8	—	1·4
1955	46·6	43·1	5	1·5	0·3	1·2	—	2·4
1956	57·5	32	4·8	1·3	0·5	1·2	0·7	2
1957	57	25	7·9	3·1	0·8	0·9	0·9	4·4
1958	59	20	10·4	3·9	1·1	1·2	1	3·4
1959	47	31	11	3·1	1·8	1·4	1·1	3·6
1960	50	23	16	3·4	1·8	1·6	1·4	2·8
1961	46	22	19	4·8	1·4	0·8	3	3
1962	38	22	19·5	6	1·5	0·5	2·3	10·2
1963	38	19	21	6	1·8	0·4	1·4	12·4
1964	43	20	24	4	1·8	0·4	2	4·8
1965	38	16	27	4	2	0·4	1·6	11
1966	39	16	24	3·7	2·2	0·3	2·2	12·6

*Includes fresh pineapples, pineapple juice and tinned pineapples.
† Includes manganese and diamonds.

Source: *Direction du Port d'Abidjan.*

Table 7.2

Value of imports at Abidjan in 1953 and 1966: percentage distribution between major sectors

Year	Foodstuffs	Non-alimentary consumer goods	Industrial materials	Equipment
1953	22·8	27·5	22·7	27·0
1966	20·0	21·0	24·0	35·0

Table 7.2 shows that imports of foodstuffs remain very important and that their proportion of the total value of goods imported has declined only slightly since 1953. The main items involved are wheat and rice, livestock, fish, sugar and salt. The growth of the plantation economy of the coastal zone has meant increased consumption of imported foodstuffs in this area in substitution for local root crops. A vigorous campaign in favour of intensive rice cultivation is beginning to give good results, but the expansion of livestock rearing is proceeding much more slowly. Imports of non-alimentary consumer goods are declining in proportion to other sectors, largely as a result of the growth of import-substitute industries in Abidjan since 1960 (particularly textiles, hosiery, oil, soap, tobacco and furniture). In contrast, the demand for industrial materials continues to increase; in this category are included sources of power, products of extractive industry and of heavy industry unknown in Ivory Coast (such as steel), and a variety of basic chemical products, machinery and spare parts. Growth of imports in the category of equipment has also been considerable, since inevitably a developing country must import construction materials (cement, ferro-concrete, paints, etc.), vehicles and electrical apparatus until these can be produced locally.

Distribution of trade by countries and monetary zones

Table 7.3 indicates the important place held in 1953 by the franc zone in the trade of Abidjan, in the context of the colonial politico-economic system then in force. France alone purchased more than half the coffee and one-third of the cocoa exported, whilst the French Union (especially North Africa) absorbed 15 per cent of Ivory Coast shipments. Over 70 per cent of imported goods came from France, whilst North Africa supplied fresh vegetables, preserved fruits, tobacco, wine, salt and cement. The trade balance of Ivory Coast with other territories within the French Union has always been favourable – an unusual situation amongst former French colonies. The dollar zone, represented chiefly by the U.S.A., is a major area of consumption for coffee and cocoa from Ivory Coast. In view of the modest volume of imports from the dollar zone, and in spite of important quantities of Venezuelan oil, the trade balance with the dollar zone has been consistently in favour of Ivory Coast since 1954. The situation in respect of the sterling zone is the reverse: Britain purchases only small quantities of timber and cocoa from Ivory Coast, since most of her requirements in tropical products are supplied by Commonwealth countries, but Britain supplies Ivory Coast with tea, textiles and machinery. As far as other countries are concerned, imports from Ivory Coast include

Table 7.3

Traffic of the port of Abidjan in 1953 and 1966, by commodities and by countries and monetary zones (percentages)

Exports By commodities:	1953	1966	Imports By commodities:	1953	1966
Coffee	47·0	39·0	Equipment	27·0	35·0
Cocoa	41·3	16·0	Industrial materials	22·7	24·0
Timber	4·8	27·0	Foodstuffs	22·8	20·0
Bananas	1·9	3·7	Non-alimentary consumer	27·5	21·0
Pineapples	—	2·3	goods		
Minerals	—	2·2			
Miscellaneous	5·0	9·9			
By countries and monetary zones:			By countries and monetary zones:		
France	53·0	40·0	France	70·4	53·5
Other territories within French Union (1953)	15·7	—	Other territories within French Union (1953)	8·5	—
African states within franc zone (1966)	—	6·4	African states within franc zone (1966)	—	13·5
Dollar zone	5·2	17·0	Dollar zone	5·4	8·0
Sterling zone	1·8	4·6	Sterling zone	7·3	6·0
Other monetary zones	24·3	10·0	Other monetary zones	8·4	7·5
E.E.C. (1966)	—	22·0	E.E.C. (1966)	—	11·5

Source: *Direction du Port d'Abidjan.*

considerable quantities of cocoa (Netherlands), timber (West Germany), bananas (Italy), coffee (Belgium, Italy), and exports include rice and beers (Netherlands), cement, textiles, tools and scientific equipment (West Germany).

Since independence, Ivory Coast has succeeded in its efforts to diversify its commercial links with a wide range of countries and to strengthen ties with stable monetary zones. Although the franc zone remains in a dominant place in terms of exports from Ivory Coast, there has been a clear decline in the relative value of these shipments (from 68·7 per cent in 1953 to 61 per cent in 1962, 44 per cent in 1964 and 46 per cent in 1966), showing that Ivory Coast has increased its range of markets. The position of the United States and other dollar-zone countries has become more important (1966, 17 per cent), whilst exports to Common Market countries in Europe (especially Italy, West Germany and the Netherlands) have significantly increased. Japan, Britain and African franc-zone countries (especially Senegal, Algeria and Morocco) are also important customers for Ivory Coast products, and sales to Central Europe and the Middle East are increasing.

In terms of imports the franc zone has preserved its predominant role to a greater extent; from 78·9 per cent of the value of goods imported in 1953, the proportion supplied by the franc zone has declined only to 73 per cent in 1962 and to 67 per cent in 1966. North Africa now supplies Ivory Coast with a greater volume of goods than France, as a result of oil cargoes. Increased imports of industrial materials and equipment have recently created an adverse trade balance with France and the franc zone, but with other monetary zones the trade balance remains largely favourable. In general terms, the range of countries from which imports are derived is narrower than that to which exports are despatched. However, imports from Eastern Europe are increasing (e.g. textiles from Hungary) and Scandinavia too is beginning to sell office equipment, refrigerators and timber-cutting machinery to Ivory Coast.

Trade balance

A major fact which emerges from this discussion is that between 1947 and 1967 the trade balance of Ivory Coast has always been favourable. Until 1958 Ivory Coast occupied a unique position within the French West African federation. Table 7.4 shows that the value of the trade surplus varied from 6 to 21 per cent between 1947 and 1958, so that each

Table 7.4

Value of the external trade of Ivory Coast shown as a percentage of the total trade of former French West Africa, 1947–58

Year	Exports	Imports	Surplus	Year	Exports	Imports	Surplus
1947	26	15	6	1953	41	23	21
1948	28	20	14	1954	47	29	16
1949	37	22	15	1955	48	28	15
1950	44	25	12	1956	44	28	17
1951	45	25	6	1957	40	25	11
1952	47	25	9	1958	43	26	7

Source: *Direction du Port d'Abidjan.*

year Ivory Coast was able to reduce the trade deficit of French West Africa as a whole. Since 1958 the trade balance has continued to be favourable, but in recent years the surplus has tended to decrease (1966, 7 per cent). This shows that up to the present time increasing imports have always been exceeded by exports which have maintained a slightly higher level, clearly a sign of a fairly healthy economic situation during the period concerned. It is likely, however, that the growth of exports of primary products will begin to level out in the near future, partly as a result of adverse world market conditions. The trade surplus hitherto maintained may therefore decrease still further and eventually disappear.

Since its opening in 1950 the port of Abidjan has developed to the full its role as the economic lung of Ivory Coast and has allowed the country to exploit resources previously held in check. Plans for the continuation of this process include the establishment of a new port at San Pedro, which would facilitate the economic development of the south-western part of the country. However, a seaport is in one sense merely an economic instrument in the service of a development policy, and the nature and pattern of economic growth reflected by the structure of external trade pose qualitative rather than quantitative problems. Some of these problems transcend the state environment and belong to West Africa as a whole; thus the development of a communal basic industrial structure which would supply those countries which have already developed light industries would reduce the cost of semi-finished products; and the regional, as opposed to the national, organisation of African economic development would allow local industries to expand through the acquisition of new markets. Other problems, in contrast, stem from the economic situation of Ivory Coast itself: amongst these are the role of foreign capital within a more independent economy; the problematical evolution from extensive agriculture to intensive cultivation, both in terms of plantation crops and food crops; and the problem of the education of Ivory Coast cultivators in the economics of modern land management. In this broad context the port of Abidjan stands as a vital element in the economic infrastructure of Ivory Coast, and demonstrates that a port with a supporting system of land communications provides an essential precondition for economic growth. Neither a modern seaport nor a wider transport system is in itself an adequate stimulus for continued growth, but the provision of transport allows (and indeed favours) development when other circumstances encourage the transition from the subsistence to the cash economy.

8. Port Development and Economic Growth: the Case of Ghana

D. Hilling

Lecturer in Geography at Bedford College, University of London, from 1961 to 1966 Mr Hilling was on the staff of the University of Ghana. He has published a number of papers on ports and economic development in West and Equatorial Africa.

For many of the developing countries of the tropical world a history of colonial dependence has combined with a geography which makes for complementarity of natural resources to ensure that their trade is overwhelmingly directed to overseas countries. This being so, the seaport becomes a major determinant of the rate of economic growth and the stage of economic development in the hinterland becomes a function of the capacity and degree of sophistication of the port facilities. This essay seeks to demonstrate, without quantifying in precise terms, the relationship that has existed between port development and economic growth in Ghana since 1945.[1]

In 1937 Ghana's total seaborne trade amounted to 1·5 million tons and was shared by eight ports of entry. Two-thirds of the total traffic and four-fifths of the exports passed through the country's only deep-water port at Takoradi, the remainder being handled at seven surf ports of which Accra was by far the most important.[2] The effects of the Second World War were profound. As the war progressed West Africa assumed a critical strategic role as a route to North Africa and the Middle East. Ships lying at anchor as much as two miles from the shore and handling cargo slowly by way of surf boats were obvious targets for enemy action and, with the exception of Accra, which continued to operate, but with much reduced traffic,[3] the surf ports were closed. Takoradi was therefore forced to handle the bulk of the normal trade plus all the military traffic

---

1 *For convenience the term Ghana will be used throughout, although more accurately the territory was the Gold Coast until it became independent in 1957.*
2 *The other surf ports were Half Assini, Cape Coast, Saltpond, Winneba, Ada and Keta.*
3 *In 1938, 298 vessels were cleared at Accra; in 1943 and 1944 the number was 38 and 30 respectively.*

and, while the country's total seaborne trade hardly changed, Takoradi's share increased to 95 per cent. Just as Ghana had started to exploit her manganese resources during the First World War, so the Second World War led to the establishment of bauxite mining and the provision of a bulk handling plant at Takoradi. The country's timber industry was greatly expanded,[1] but a general lack of imported goods led to a steady deterioration of the country's infrastructure.

Ghana emerged from the war with a relatively healthy export trade but with an economy gravely deprived of capital and consumer goods. During the late 1930s there were indications of an increasing awareness, both in London and Ghana, of the need to diversify the territory's economic base, but it took the changed political climate of the early post-war years to give urgency to this demand. A new constitution of 1946 made provisions for an elected legislature, and the formation in 1947 of the United Gold Coast Convention and in 1949 of the Convention People's Party were major steps along the road to full internal self-government (1954) and complete political independence in 1957. These far-reaching political changes were accompanied by an increasingly aggressive approach to economic and social development and the demand for economic as well as political independence.

The insatiable demand for tropical raw materials in the immediate post-war years, combined with the need to reconstruct and develop Ghana's infrastructure, led to a great expansion of the country's external trade. Since external trade was virtually synonymous with seaborne trade,[2] the ports became a vital element in the economic development process. Even when due allowance is made for increasing prices, the leap in the value of imports from £7·6 million in 1938 to £45·4 million in 1949 is impressive. Food imports increased in value from £1 million to £6 million and consumption of tobacco, drinks and textiles increased four, seven and ten times respectively. Cement imports doubled in the same period.

There can be no doubt that higher personal consumption and more rapid development increased trade so that the physical capacity of the country was severely strained and inflation commenced. 'To say which is the most serious of these strains would be hard, but a strong case can be made out for treating port capacity as the main bottleneck.'[3] The port

---

1 *Log production increased from 14,300 cu. m. (0·5 million cu. ft) in 1939 to 77,000 cu. m. (2·7 million cu. ft.) in 1945.*

2 *In 1966 a mere 1·9 per cent of Ghana's trade was with African countries, and nearly 30 per cent of that passed through the ports.*

3 D. Seers and C. R. Ross, Report on the Financial and Physical Problems of Development in the Gold Coast (*Accra: Government Statistician, 1952*).

traffic statistics for post-war years (Table 8.1) give some support to this suggestion.

Clearly in early post-war years the ports had to handle tonnages far in excess of anything they had handled before 1939 or could reasonably be expected to handle with the facilities available. Although world trade continued to boom, there was an overall slackening in the growth rate of port traffic after 1949. The 1950 traffic figures were certainly affected by a strike which held up trade, and the rather higher totals in the following year represent a carry-over effect. The surf ports, closed during the war, were increasingly brought back into use, in spite of their obvious deficiencies. While in pre-war and immediate post-war years there was a marked imbalance between import and export tonnages, the gap was gradually narrowed.

The greater volume and range of imports made new demands on port facilities, took longer to handle than the homogeneous bulk exports, and led to port congestion and shipping delays.[1] Mineral exports were loaded at special bulk-handling berths, capable of dealing with tonnages well in excess of demand. Bagged cocoa, the main export in terms of value, could be handled even at the surf ports in considerable tonnages, and although the methods were far from efficient they proved adequate as long as the beaches remained uncongested by imported goods. Timber, exported through Takoradi, created problems, but to some extent these were due to shortage of shipping space rather than inadequate port facilities.[2] Export tonnages depended on the slow growth of productive capacity in mining, agriculture and forestry, and increased production had to await improvements in the neglected infrastructure which in turn depended on imports. Thus, while exports advanced at a slow rate, the import tonnages increased rapidly and presented the ports with their main problem.

The country's ports were clearly not designed to accommodate traffic of this magnitude and varied character. Deep-water facilities existed only at Takoradi, where an artificial harbour had been opened to traffic in 1928. The port provided special berths for loading manganese and bauxite, a coaling berth and tanker mooring, but only two general cargo berths for imports. The northern and western wharves could accommodate lighters, and most of the timber and cocoa was rafted and lightered from these wharves to ships at moorings in the shelter of the breakwaters. Although

1 *United Africa Company, 'Port Capacity and Shipping Turnround in West Africa'*,
   Statistical and Economic Review, *vol. 19 (1957)*.
2 *C. Leubuscher*, The West African Shipping Trade, 1909–1959 (*Leyden, 1963*).

Table 8.1
Ghana's port traffic, 1938 and 1945–67
(1,000 tons)

	TAKORADI		ACCRA		TEMA		OTHER		COUNTRY Imports		Exports		Total	
	Imports	Exports	Imports	Exports	Imports	Exports	Imports	Exports	Tons	% change	Tons	% change	Tons	% change
1938	299	477	92	129	—	—	40	54	431	—	653	—	1,084	—
1945	282	1,135	62	99	—	—	—	—	344	−20.2	1,234	88.9	1,578	45.6
1946	320	1,206	93	91	—	—	1	4	414	20.3	1,301	5.4	1,715	8.7
1947	355	986	156	58	—	—	7	15	518	25.1	1,059	−18.6	1,577	−8.4
1948	444	1,179	184	56	—	—	18	14	646	24.7	1,249	17.4	1,895	20.2
1949	556	1,319	273	65	—	—	35	25	864	18.9	1,409	8.7	2,273	20.10
1950	583	1,280	245	66	—	—	51	27	881	1.9	1,373	−2.6	2,254	−0.84
1951	675	1,378	320	61	—	—	53	25	1,048	18.9	1,464	6.6	2,512	14.4
1952	711	1,317	312	55	—	—	33	27	1,056	0.76	1,399	−4.4	2,455	−2.3
1953	717	1,403	390	69	—	—	51	31	1,158	9.7	1,503	7.4	2,661	8.4
1954	719	1,157	387	60	—	—	54	28	1,160	0.17	1,245	−17.2	2,405	−9.6
1955	854	1,272	512	69	—	—	60	30	1,426	22.9	1,371	10.1	2,797	16.3
1956	799	1,529	529	78	—	—	56	28	1,384	−2.9	1,635	19.3	3,019	7.9
1957	874	1,679	543	80	—	—	55	35	1,472	6.4	1,794	9.7	3,266	8.2
1958	764	1,525	583	65	—	—	50	27	1,397	−5.1	1,617	−9.9	3,014	−7.7
1959	815	1,720	675	81	—	—	55	29	1,545	10.6	1,830	13.2	3,375	12.0
1960	1,007	1,895	791	89	—	—	45	51	1,843	19.2	2,035	11.2	3,878	14.9
1961	1,107	1,558	877	155	172	35	57	38	2,213	20.1	1,786	−12.2	3,999	3.1
1962	736	1,562	206	—	622	206	17	7	1,581	−28.6	1,775	−0.6	3,356	−16.1
1963	717	1,504	—	—	1,010	291	—	—	1,727	9.2	1,795	1.1	3,522	4.9
1964	448	1,513	—	—	1,699	578	—	—	2,147	24.3	2,091	16.5	4,238	20.3
1965	382	1,643	—	—	1,537	658	—	—	1,919	−10.6	2,301	10.0	4,220	−0.4
1966	417	1,597	—	—	1,357	584	—	—	1,774	−7.6	2,185	−5.2	3,955	−6.3
1967	242	1,436	—	—	1,510	558	—	—	1,752	−1.2	1,994	−8.7	3,746	−5.2

working barely to capacity in pre-war years, a report of 1937[1] had suggested an extension of the main wharf by 230 m. (750 ft), the provision of more open storage, the improvement of the rail system and additional quay cranes. None of this work had in fact been started in 1947 when the consulting engineers were asked to make a fresh appraisal of the situation.[2]

At Accra a 183-m. (600-ft)-long breakwater gave some protection for small craft and the beach landing in its lee. A jetty with 3-ton travelling steam cranes had no more than 2 m. (7 ft) of water at its head at low water and could be used only by lighters or surf boats (Plate 7). All produce for export was loaded directly into surf boats on the beach, but while small packaged imports of up to 0·3 ton could be manhandled at the beach, larger items were off-loaded at the jetty. With a capacity of 1½ to 2 tons per journey, or up to 4 tons if two surf boats were lashed together, the method was clearly inefficient. Cargo was placed on gratings in the surf boat and covered with tarpaulin, but damage by sea-water was still frequent and packages easily lost overboard. Marine insurance rates were understandably high. While the nature of the goods that can be handled in this way is clearly restricted, there is in theory no limit to the volume that can be dealt with. In practice, however, the number and distance from the shore of safe anchorages, the number of surf boats available, the landing space on the beach and the access to and capacity of warehouses will limit capacity. Accra demonstrated the difficulty of assessing the capacity of such ports. In 1950[3] it was said that under ideal conditions the port could handle as much as 40,000 tons a month but that a more realistic annual traffic was 295,000 tons. In fact in 1961 Accra handled over one million tons,[4] the bulk of which was general imports and cement.

Before the war, cargo had been handled at six other surf ports: Half Assini, Cape Coast, Saltpond, Winneba, Ada and Keta (Fig. 8.1). In 1946 and 1947 Cape Coast, Winneba and Keta were brought back into use in response to the increased demand for port capacity. Cape Coast and Winneba both have limited beach space and Keta, while it has an extensive beach, lacks shelter at the landing place. Their capacity was therefore very

1 *Rendel, Palmer and Tritton*, Report on Proposed Improvements at Accra (*London, 1937*).
2 *Rendel, Palmer and Tritton*, Report on Proposed Extensions at Takoradi Harbour (*London, 1947*).
3 *United Africa Company*, '*Surf Port Operations on the Gold Coast*', Statistical and Economic Review, *vol. 5* (*1950*).
4 *Central Bureau of Statistics*, Economic Survey 1961 (*Accra, 1962*).

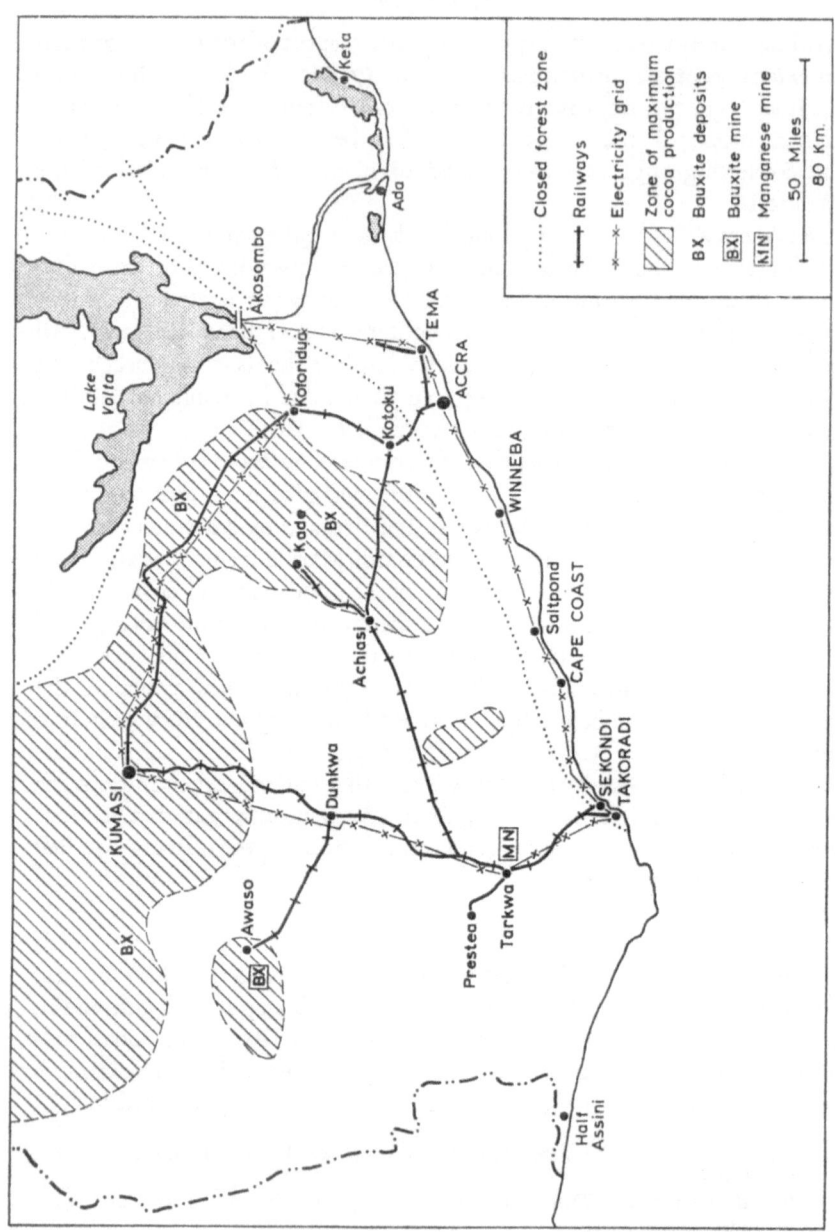

Fig. 8.1 Economic map of southern Ghana

limited and they could be looked upon only as supplementary to the main ports.[1]

'Port capacity' is a term frequently used but rarely defined. Indeed, the variables are so numerous that exact definition is virtually impossible and comparison between ports on this basis of no great value. The concept of 'optimum capacity' is more useful and might be defined as the point at which a port increases its traffic only at the expense of congestion, delays to shipping and generally increased costs. There is no doubt that the rising freight rates on the West African range in the years after the war were in large measure the result of increased turnround time. Before the war, ships operated strict schedules and 'loss of time awaiting a berth was virtually unknown',[2] but in the first six months of 1956 vessels of the West African Lines Conference alone lost 121 days awaiting cargo handling at Takoradi and Accra. In 1954 and 1955 approximately 70 per cent of the recorded delays to Palm Line vessels were attributed to inadequacies at the ports.[3] There were examples from Accra of vessels being at anchor for as much as sixteen days in order to discharge 158 tons of freight.[4]

The capacity of a port is obviously related to the capacity of the inland transport system serving it. Road access to the two main ports was limited and the single-line rail system inadequate. Marshalling facilities were restricted, and the pile-up of goods at Accra and Takoradi added to the problems of collection and forwarding. The eastern part of the country, with no deep-water port, had perforce to receive most of its bulky imports by way of Takoradi and the railway via Kumasi, there being no direct Accra–Takoradi rail link until 1956. All coal for railway storage dumps had to follow the same route.

In 1950 the ports handled 180,000 tons of cement, and the Conference proposed a limit of 12,000 tons a month in view of the delays experienced by its ships. In the same year the estimated demand for cement to meet development commitments was 25,000 tons a month.[5] The ports were operating far beyond their optimum capacity and this was inevitably reflected in both the rate and cost of development. Unless port facilities were both improved and expanded, the country's economic growth would be severely checked.

As early as 1947 the consulting engineers expressed the view that

1 *Early in 1961, with severe congestion at Accra, the author's goods were off-loaded at Winneba!*
2 *United Africa Company, op. cit.*
3 *Ibid.*
4 *Port of Accra*, Monthly Reports (*Accra, 1960*).
5 *Seers and Ross, op. cit.*

facilities at Takoradi were 'insufficient for present and anticipated future trade'.[1] The bulk mineral loaders for manganese and bauxite were considered adequate, but the stern-on mooring of tankers and the discharge of oil by way of pipelines floated alongside on pontoons was clearly insufficient. The handling of the rapidly expanding timber exports and general cargo imports were identified as the major areas for improvement, together with the necessity for more extensive flat land for storage, access roads and railway sidings.

The existing facilities for timber consisted of 240 m. (800 ft) of shallow-water quay in the western part of the harbour. Logs were stored and rafted under the main breakwater and ships handled logs directly from the water with their own gear. Sawn timber was conveyed by lighter either direct from rail trucks or after storage on the quay. Until the later 1940s the quantity of sawn timber hardly justified more elaborate facilities, but the trade increased from 3,892 cu. m. (140,377 cu. ft) in 1945 to 43,416 cu. m. (2,057,114 cu. ft) in 1950 and the facilities became completely inadequate. Total storage of logs amounted to 6,000 tons, and it was thought desirable to maintain stocks of 30,000 tons to ensure steady export traffic in the face of intermittent supplies from up country. The consultants proposed reclamation in front of the existing wharf and the construction of three shallow-water docks (Fig. 8.2), each of 152 m. (500 ft) in length, with a total quayage of 990 m. (3,250 ft) and storage for 22,000 tons of logs and two ten-bay sheds with overhead gantry cranes for handling sawn timber.

To accommodate general cargo it was proposed to extend the main quay to the end of the lee breakwater so providing three additional berths. The seaward berth was intended for coal handling,[2] and the extension meant the removal of the bauxite berth from the inside to the outside of the breakwater. The effect of this was to more than double the port's general cargo capacity. A new tanker berth consisting of well-fendered breasting blocks was proposed for the north side of the lee breakwater.

If handling capacity was to be effectively increased, landward developments were as important as additional quayage. The harbour area is backed to the west by a small bluff which severely restricts growth, and to the north Cox's Fort Hill provided a bottleneck for the railway. The consultants proposed the levelling of the hill and the use of the material for infilling on a large area of foreshore reclamation and on the timber

---

1 *Rendel, Palmer and Tritton*, Report on Takoradi Harbour, *op. cit.*
2 *This never became necessary because the railway, the main user of coal, gradually changed over to diesel traction.*

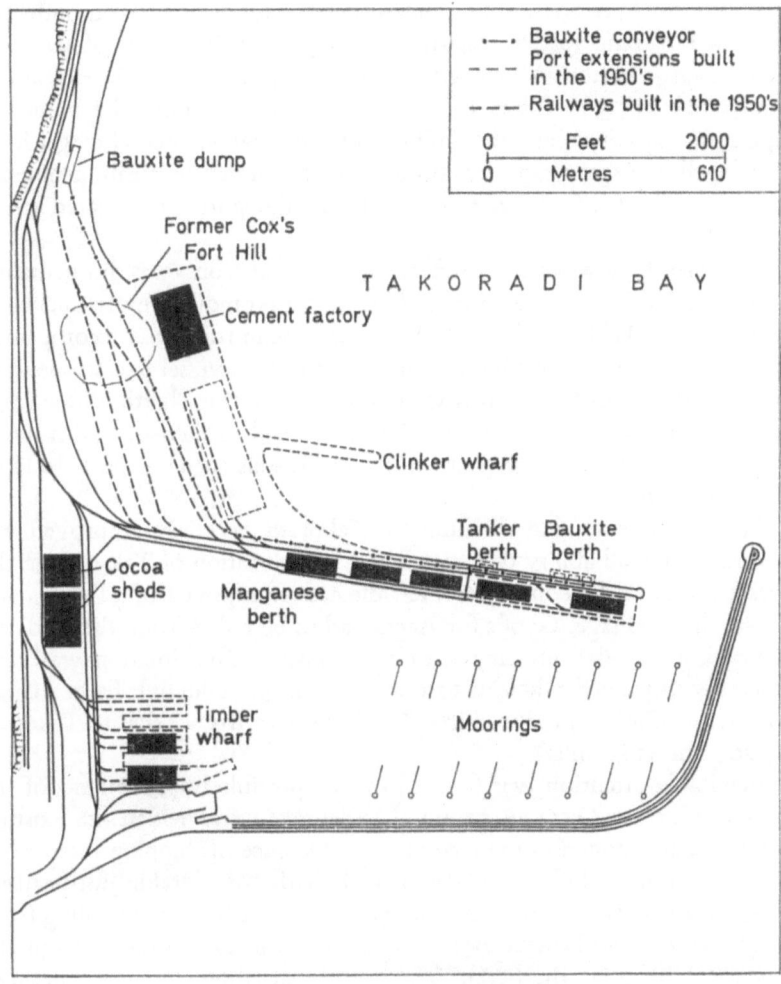

Fig. 8.2   The port of Takoradi

wharf reclamation. On the level land so acquired a new system of railway marshalling yards, cargo-handling platforms and open storage could be provided.

The contracts were awarded in June 1949 and work started almost immediately. The main problem was the phasing of the work in such a way that disruption of normal traffic was kept to a minimum. The number of vessels cleared and the import tonnages did in fact increase steadily while construction work proceeded. The new timber wharf came into operation in May 1951, and after a total shutdown of seven months the relocated bauxite loader restarted in May 1952. The first of the new general cargo berths was opened to traffic in April 1953 and the extensions completed by 1955 (Plate 8). With these new facilities the country's optimum capacity reached 3 million tons, but while general cargo for a brief spell had spare capacity, timber still continued to create problems. In 1955 a stock of 35,000 tons of logs in the port area created grave congestion.[1]

In general, however, the effect of the extensions was immediately apparent. Table 8.1 showed that after 1945 Takoradi's imports increased very rapidly. Table 8.2 shows that this increase in trade was accompanied until 1954–5 by a steady increase in the time each vessel had to spend in port, most of which time was wasted in waiting for a berth. After 1955, while the number of vessels entering continued to increase and imports and exports reached new peaks, the time vessels spent in port declined very markedly.

While the extended facilities at Takoradi certainly improved the country's overall ability to expand trade, the location of Takoradi in the western part of the country still left the eastern region around Accra at a severe disadvantage. Goods for Accra had to be railed from Takoradi via Kumasi, a two-day journey with a night stop. The situation was considerably improved when in 1956 an 82-km. (51-mile) link from Achiasi to Kotoku (Fig. 8.1) reduced the overall distance from Accra to Takoradi by 265 km. (165 miles).

Inevitably attention was focused on the possible improvement of the port facilities at Accra. A proposal to lengthen and widen the existing jetty was abandoned in the face of the reluctance of shipping interests to reintroduce large lighters. They argued, with considerable justification, that port capacity would be increased only marginally since handling from large lighters would have been slower than the manhandling from the small surf boats on the beach. The traditional labour-intensive methods had much to recommend them except for handling bulky items. Minor improvements were effected on the beach, but the consultants concluded that the needs of the country could best be met by 'further extension of Takoradi Harbour, combined with planned developments of the road and rail communications radiating from that port'.[2] It is doubtful if this was a realistic appraisal of trends in traffic even in 1951.

1 *Railway and Harbour Administration*, Report, 1954/5 (*Accra, 1955*).
2 *Rendel, Palmer and Tritton*, Report on Proposed Improvements at Accra Harbour (*London, 1951*).

The steady expansion of the country's economy and the favourable balance of trade created a large potential demand for imports. Social and

Table 8.2

Takoradi port traffic, 1937–8 to 1963–4

	Vessels entered	Average vessels in port daily	Average port time per vessel
1937–8	752	4·18	19·51
1945–6	618	4·93	57·03
1946–7	652	6·14	88·49
1947–8	736	7·11	68·11
1948–9	893	8·90	67·27
1949–50	932	9·15	75·69
1950–1	961	9·82	78·92
1951–2	987	9·16	78·76
1952–3	970	9·42	85·03
1953–4	985	9·53	80·28
1954–5	989	10·95	96·40
1955–6	1,013	10·23	81·20
1956–7	1,012	9·81	73·66
1957–8	1,253	9·84	59·14
1958–9	1,508	11·03	64·42
1959–60	1,557	11·32	65·12
1960–1	1,630	12·21	61·43
1961–2	1,586	8·36	41·74
1962–3	1,429	10·09	54·21
1963–4	1,319	9·54	54·73

Source: *Railway and Harbour Administration* Annual Report.

economic stability required that such a demand should be met. In 1951 the optimum capacity of the ports was of the order of $2\frac{1}{2}$ million tons, and the Takoradi extensions increased this to at least 3 million tons per annum, but trade projections even then suggested totals of 3·8 to 4 million tons by the later half of the 1950s.[1] Clearly growth would be held in check unless further facilities could be provided. Even so, it is debatable whether or not the facilities would have taken the form of a new artificial harbour had it not been for decisive action on the question of harnessing the River Volta for hydro-electric power.

As early as 1925 there had been a proposal to smelt Ghana's bauxite, utilising power generated by the Volta river, and such a scheme received intermittent attention until 1939. Renewed activity after the war led to a favourable appraisal of the scheme by government-appointed consultants in 1951[2] and the setting-up of the Volta River Preparatory Commission, whose report[3] firmly established the economic and technical feasibility of the project and the main outlines of the work involved. The whole scheme was clearly inconceivable without deep-water port facilities, and a new port had to be an integral part of the Volta River Project. However, the 1951 report concluded that 'the provision of a deep-water port is, in our view, essential to the future of the Colony, irrespective of whether the power scheme is carried out or not'.[4]

The demand for and the location of new port facilities was from the outset influenced by both the requirements of the general economic development of the territory and more specifically by the logistics of the Volta River Project. These influences combined to suggest a location in the eastern part of the country. Takoradi adequately served the west of the country, but even after the completion of the Achiasi–Kotoku rail link transport services to the east were unable to support substantial traffic flows. The Eastern Region is the most densely peopled region of the country, and Accra, the capital, provided a major growth pole. Between 1948 and 1960 the Accra Capital district more than doubled its population from 222,000 to 492,000, and the average annual increase of population in Accra and the Eastern Region was respectively 10·16 and 5·33 per cent.[5] Accra, with its rail links to Kumasi and its developed administrative and commercial functions, provided the obvious coastal focus in the east.

The locational considerations of the Volta River Project have been

1 Sir William Halcrow and Partners, *Report on the development of the River Volta Basin* (London, 1951).
2 Ibid.
3 *Volta River Preparatory Commission*, Report (London, 1956).
4 Sir William Halcrow and Partners, op. cit.
5 T. Hilton, '*The Population of the Upper Region of Ghana*', in J. C. Caldwell and C. Okonjo, The Population of Tropical Africa (London, 1968).

described in detail elsewhere,[1] but a proposal to use bauxite from Kibi, relatively close to the Accra–Kumasi railway and the location of the power source at Akosombo where the Volta cuts through the Akwapim Togo Range, clearly indicated a port in the region eastwards from Accra. Accra itself was clearly incapable of development to fulfil this function. There was a long history of siltation in Accra Bay, and the landward approaches to the port were congested by dense urban development. The privately prepared scheme for a Volta river project by West African Aluminium Ltd[2] suggested a smelter close to the dam and reliance on water transport for bulk movements. Their scheme involved the construction of training moles at the mouth of the Volta, the directing of river flow and dredging of the river's western arm to provide the harbour basin, and the improvement of the Volta to become the main link with the smelter at Kpong. Dr Ringers of The Hague and Professor Thijsse of Delft supported the idea of a port of Ada, and concluded that 'there is no choice, the harbour has to be made at the mouth of the Volta'.[3]

The decision of government consultants (Sir William Halcrow and Partners) to consider a smelter on the coast and a later decision to operate initially on imported alumina meant that Ada ceased to have any advantages over a site nearer Accra. Halcrow accepted that a port could have been built at Ada, but submitted that the liability of flooding, the instability of the river mouth and the reduced effect of scour upon completion of the dam would render the port more difficult and costly to construct and maintain than had been estimated. Halcrow directed their attention to two sites closer to Accra, Teshie and Tema, and demonstrated the topographic and hydrographical advantages of the latter. Being only 24 km. (15 miles) east of Accra, Tema could easily be connected to the existing rail and road network and could effectively develop as the capital's outport. Ada, 112 km. (70 miles) from Accra, could never have been a satisfactory outport. Thus, unless Ada could be demonstrated to have overwhelming advantages as a site for the port to serve the Volta River Project, the advantages of Tema to serve general economic requirements 'lead to the inevitable conclusion that the combined port should be sited at Tema'.[4]

At Tema, deep water close to the shore and a steeply inclined rocky sea

1 D. Hilling, 'Ghana's Aluminium Industry: Some Locational Considerations', Tidjschift voor Economische en Sociale Geografie, vol. 55, no. 5 (1964).
2 C. St J. Bird, The Volta River Scheme, Report to West African Aluminium Ltd (1949).
3 J. Th. Thijsse, communication to Messrs Coode and Partners (3 Mar 1952).
4 Sir William Halcrow and Partners, op. cit.

bed minimised dredging. A small rocky headland provided a 'root' for
the main breakwater which, once started, would provide shelter for
vessels unloading constructional material. The site was well located with
respect to the Shai Hills, where rock suitable for breakwater construction
was available in large quantities and easily quarried. On these counts the
Tema site had distinct advantages over Ada, and the Government accepted
the consultant's proposals for the new port in spite of the arguments
advanced by the inhabitants of the Volta Region, who saw in the port at
Ada a long-needed stimulus to regional development. Preliminary work
started at Tema in 1951, the main port works started in 1954 and the port
was opened for restricted commercial operations in January 1961 (Fig. 8.3).
In 1962 the port became fully operative, but even as work continued
extensions to the original scheme were approved and incorporated.

In the meantime, however, the additional facilities at Takoradi and the
seemingly infinite capacity of Accra's primitive facilities permitted further
expansion of the economy. In terms of tonnage handled there was an
increase from 2·4 million in 1954 to 4 million in 1961. With 60 per cent of
the country's income derived from cocoa exports, Ghana's level of
economic activity has always been dangerously dependent on the size of
the cocoa harvest and the prevailing world prices. Despite the fluctuations
of yield inevitable with an agricultural crop, the general trend of production was upwards during the 1950s, and with world cocoa prices at a
reasonably high level personal incomes were high. The value of total
exports increased from 242 million cedis (₡2·45 = £1) in 1955 to
₡295 million in 1960, and over the same period private consumption
expenditure increased from ₡619 million to ₡833 million.[1] Superimposed
on this was a marked increase in government spending, from ₡62 million in
1955 to ₡115 million in 1960. Much of this expenditure was on social
and economic infrasturcture. In the 1960s further increases in the volume
of cocoa exported were accompanied by a sharp drop in prices, and by
1965 the value of exports had only increased to ₡300 million. There was a
general slowing down in the rate of economic growth.

There can be no doubt that during the 1950s the limitations on growth
were largely of a physical nature, with port capacity a major element. By
1961, however, it could be claimed that the infrastructure was 'adequate
for the present levels of economic development',[2] and during the 1960s
the main restraints on growth have been financial, with a marked slowing
of the rate of growth from an average of 6·9 per cent per annum from

1 Africa Research Bulletin, *vol. 5, no. 6 (1968)*.
2 *Central Bureau of Statistics*, Economic Survey, 1961.

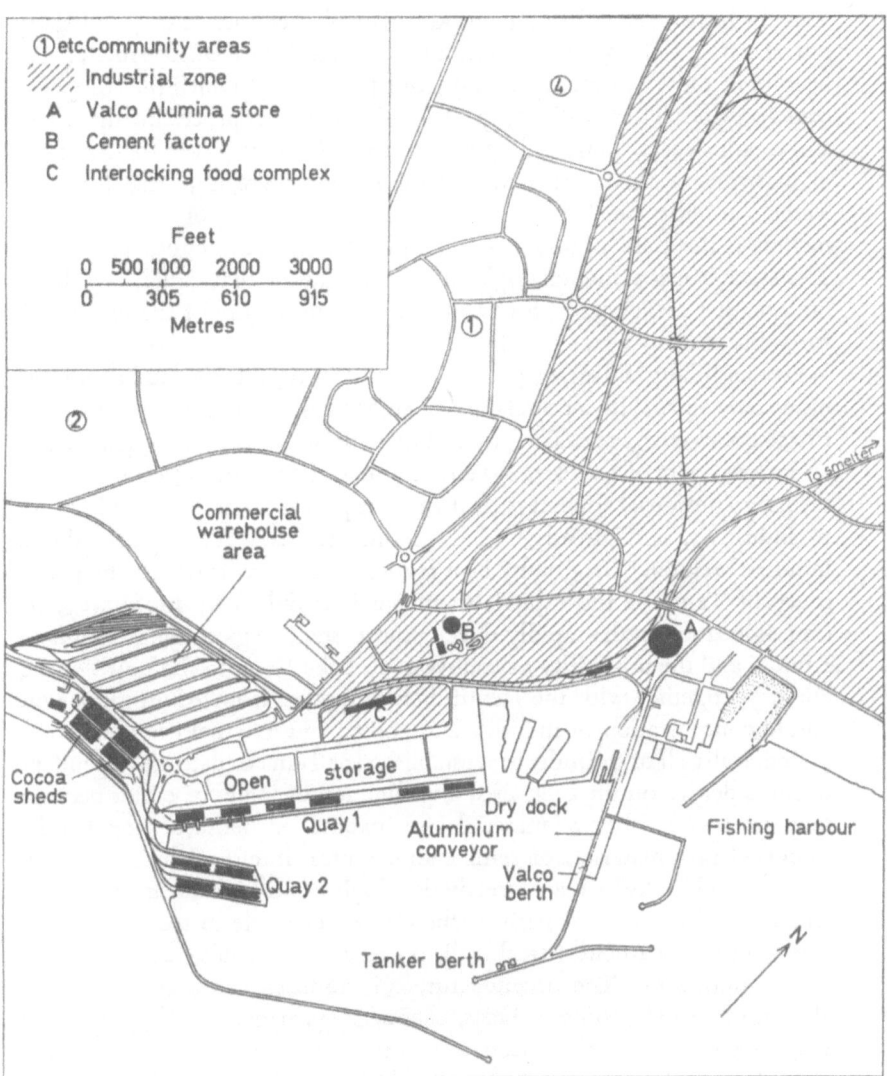

① etc. Community areas
⧲ Industrial zone
A Valco Alumina store
B Cement factory
C Interlocking food complex

Feet
0 500 1000 2000   3000
0   305   610   915
Metres

Fig. 8.3 The port of Tema

1955 to 1960 to 3 per cent from 1961 to 1966. Nevertheless, whilst the rate
of growth has been slower developments have been impressive.

Work on the Volta river dam and generators was completed in 1966,
well on schedule and at less than estimated cost, a feat which could

scarcely have been achieved without Tema's facilities and an efficient infrastructure. The Volta Aluminium Company's (VALCO) smelter has been built only $2\frac{1}{2}$ km. ($1\frac{1}{2}$ miles) from the port, at which the company has its own berth for the discharge of alumina and other processing materials and to which it is connected by private road (Plate 9). Over 500,000 tons of material for the construction of the smelter were imported through Tema, and the smelter generates over 300,000 tons of trade a year, including the bulk of the 105,000 tons of ingot aluminium produced. The smelter is a prime example of a port industry that could have been built neither at Takoradi, for want of space, nor at any of the surf ports, for want of deep-water handling facilities.

From a fishing village of 4,000 inhabitants in 1954 Tema has become an industrial centre with a population of 80,000 (1968) and further expansion to 250,000 is envisaged. The physical growth of Tema is being planned as an integral part of the Accra–Tema municipal area, of which it is to be the heavy industrial zone. It is clear that the port is attracting a wide range of industries, and already it has become the country's largest single industrial region. 'Geography is in favour of the new town.'[1] The proximity of Tema to the capital, to which it is linked by motorway, its central location in the rapidly developing Eastern Region, the rail links to Kumasi and the possibility of transport on Lake Volta to the north of the country, together with the plentiful cheap power from Akosombo, will ensure concentration of industry and consequent external economies.

Practically all of the industries established at Tema to date are dependent to some degree on the bulk-breaking function of the port, either because they process local raw materials for export or because they handle imported raw materials or semi-manufactures. Industrial linkages, both forward and backward, are already developing and in this sense Tema is a true industrial complex, perhaps the clearest example in tropical Africa. The aluminium smelter, steel mill, oil refinery and shipyard will act as leader industries. The manufacture of domestic and constructional aluminium goods, roofing sheets, chemicals, pharmaceuticals, paints and soaps, cigarettes, textiles, electrical components, vehicle assembly and a variety of food-processing and service industries such as engineering and printing are logical developments either from the leader industries or at a port location. The Volta River Project certainly represents the purposeful beginning of industrialisation in Ghana, and it is at Tema that the growth has been focused, so much so in fact that the Government is now providing

1 *Doxiadis Associates*, Accra–Tema–Akosombo: Regional Programme and Plan (*Tema, 1961*).

incentives for industrialists who establish enterprises away from this zone![1] While this policy may be desirable on social and political grounds, there can be little doubt that on economic grounds industry finds at Tema an optimum location.

Takoradi has had a much less important role industrially, and although its facilities have existed for longer than those of Tema, there is no comparison in the amount of industry the port facilities have stimulated. Industrial development in the immediate vicinity of the port has been restricted by physical geography and also by urban growth. The recently established cement factory, located as it is on the land reclaimed in the 1950s, is the only example of a true 'industrial port' activity. Other industries, such as sawmilling, plywood, cigarettes, paper, chemicals and cocoa processing, are located well away from the port area, although in some cases they are dependent on the port. Takoradi lacks Tema's extensive gently undulating coastal plain, so admirably suited for large-scale industrial plant, and neither does it have the benefit of proximity to the country's political, cultural and commercial capital, Accra. However, it is well located to serve the timber, cocoa and mineral-producing regions of western Ghana and is the natural distribution centre for that area. Plans to construct a dual-carriageway or motorway-standard road from Takoradi to Kumasi[2] will greatly facilitate both forwarding and collection, and Takoradi could then recapture some of the trade which is now passing through Tema.

The surf ports have now ceased to operate, and traffic flows are focused on the two deep-water ports.[3] Like the surf ports it replaced, Tema is primarily an importing port and in 1966 handled 65 per cent of the country's dry cargo imports and 76 per cent of the total imports (Table 8.3). However, it handled a mere 13 per cent of the dry cargo exports and cocoa made up 90 per cent of this total. These figures reflect the rapid economic growth of the eastern part of the country, the development of a large non-agricultural population, the need for food imports and the relative poverty of Tema's hinterland in raw materials and exports with the exception of cocoa, 57 per cent of which passed through the port in 1966. Clearly Tema's industries, many of them potential exporters, have not started to make any great impact on the trade structure of the port.

Takoradi's import trade declined markedly with the opening of Tema (Table 8.1) and fewer ships have been calling. There has been a great

1 *Col. A. A. Afrifa, Budget speech (Accra, July 1968).*
2 Africa Research Bulletin, *loc. cit.*
3 *D. Hilling, 'Tema: the geography of a New Port',* Geography, *vol. 51, no. 2 (1966).*

Table 8.3

Trade structure of Tema and Takoradi, 1964–7

	1964 Tema	1964 Takoradi	1965 Tema	1965 Takoradi	1966 Tema	1966 Takoradi	1967 Tema	1967 Takoradi
Dry cargo exports:								
Cocoa	193,434 (84·3)	181,680 (12·1)	295,040 (91·3)	237,785 (14·4)	218,734 (90·4)	167,595 (10·4)	177,466 (77·6)	131,500 (91·1)
Total	229,461	1,499,330	316,676	1,642,175	241,848	1,597,091	228,448	1,436,950
Dry cargo imports:								
Cement	429,995 (48·0)	191,703 (43·9)	354,743 (44·2)	156,890 (42·2)	302,297 (40·3)	171,822 (42·1)	264,133 (38·7)	89,456 (36·7)
All building materials	528,494 (59·0)	215,013 (49·2)	462,331 (57·7)	183,871 (49·5)	351,048 (46·8)	194,190 (47·6)	288,664 (42·6)	114,145 (47·1)
Foodstuffs	115,911 (12·9)	47,127 (10·8)	122,903 (15·3)	49,388 (13·3)	142,078 (18·9)	71,171 (17·4)	138,343 (20·3)	42,326 (17·3)
General	181,953 (20·3)	108,917 (24·9)	239,728 (29·9)	97,629 (26·2)	216,603 (28·9)	71,829 (17·6)	185,020 (27·1)	59,254 (24·4)
Total	895,723	436,958	800,710	371,561	748,580	407,579	681,397	242,312
Total all dry cargo	1,125,184	1,936,288	1,117,386	2,013,736	990,428	2,004,670	909,845	1,679,262
Petroleum								
Imports (crude)	803,166	11,294	735,752	10,799	607,646	9,965	729,219	18,896
Exports (products)	349,153	13,462	341,654	1,290	342,096	—	336,595	4,760
Total all trade	2,277,503	1,961,044	2,194,792	2,025,625	1,940,170	2,014,635	1,975,659	1,702,918

Figures in brackets show percentages of ports' imports or exports.
Source: *Railway and Harbour Administration Reports.*

reduction in the number of hours each vessel spends in port (Table 8.2), a clear indication that the port was operating beyond its optimum capacity. By virtue of the manganese, bauxite and timber from its hinterland Takoradi has managed to keep well ahead of Tema in export tonnage, but in terms of total trade there is very little difference in the tonnages handled.

The ports are currently handling about 4 million tons of cargo a year, a total which is kept artificially low by financial restrictions on growth. When it is remembered that Takoradi and the surf port of Accra handled only marginally less than this in 1961, the present facilities must have a capacity well in excess of the tonnages being recorded at present. Thus, and possibly for the first time in its history, the country has a rational port system and capacity adequate for its needs. How long the ports will remain adequate it is difficult to say. Clearly Ghana's trade is still capable of great expansion, but the rate will depend on the world and national economic situation. At Tema there is considerable space for additional conventional berths within the harbour area. It remains to be seen whether or not general trends in world shipping, particularly increasing ship size and unitisation, will make revolutionary new demands on the ports. Hopefully, however, Ghana's ports will not, for a long time, act as a check on the country's material economic development.

F

9. Cotonou: Some Problems of Port Development in Dahomey

A. Mondjannagni

M. Alfred Mondjannagni is a member of the staff of the Institut de Géographie Tropicale in the University of Abidjan, Ivory Coast. He is co-author (with M. Assane Seck) of L'Afrique Occidentale (Paris, 1967).

In April 1965 the pier at Cotonou, through which the bulk of Dahomey's external trade had passed for almost three-quarters of a century, finally closed down and was replaced by a deep-water port, the most recent of its kind on the West African coast. Other deep-water terminals were already in service along the coast of the Gulf of Guinea: Lagos (1903), Takoradi (1928), Abidjan (1950) and Tema (1962), but the addition of Cotonou to this pattern is of interest from several viewpoints. The unique characteristics of the physical setting and the historical development of maritime transport in Dahomey provide a general framework. Technical and political problems involved in the development of the new port are of interest; and it is possible now to comment upon some of the effects of the opening of the port upon the general economic situation in Dahomey (where the growth of overseas trade in recent years has not been very encouraging), and more specifically upon the town of Cotonou, the coastal fishing communities of Lake Nokoué and the Porto Novo lagoon, and finally upon various social problems raised by the development of this new element in West Africa's transport infrastructure.

Physical characteristics of the Benin coast, with special reference to the coast of Dahomey

The coastal landscape in which the new port of Cotonou is set exhibits the same topographical and morphological features as are found generally along most of the coast of the Gulf of Benin: a succession of innumerable sand bars, of varying size, lagoons forming a line parallel to the coast, and mud flats flooded at high water and linked to the deltaic areas. The generally rectilinear sand bars are interrupted by the Boca del Rio channel, the Grand Popo channel and the channel leading to Cotonou lagoon; these openings allow interchange between the freshwater lagoons and the open sea, and the periodic entry of sea-water into Lake Nokoué has considerably modified the biological environment.[1] The Cotonou channel has, however, tended to become blocked annually by sand carried by the coastal currents, chiefly during the dry season when the level of water is low; during the wet season the sand barrier is breached by the flood waters or by the local people anxious to minimise inundation. Both older and more recent sand bars may be distinguished: the older examples, generally very extensive in an east–west direction, provide important settlement sites; lower in altitude than the more recent bars, they are occasionally broken up into circular or oval mounds.

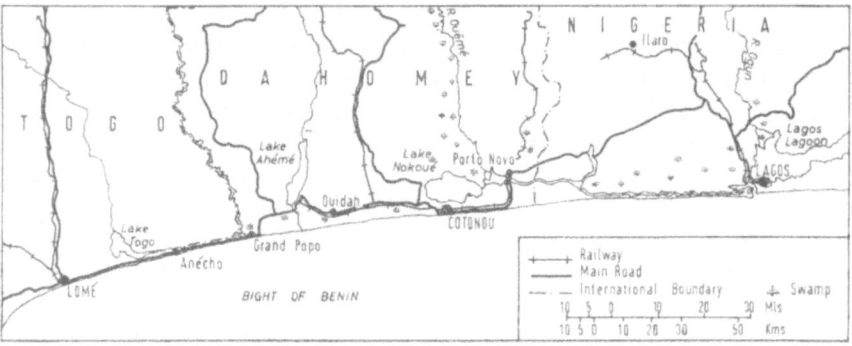

Fig. 9.1   Location of the port of Cotonou

The lagoons of Porto Novo, Ouidah and Lake Nokoué (Fig. 9.1) are formed by drowned valleys or rias, aligned parallel to the present coast and constituting an important means of communication between the various coastal regions and communities of the Gulf of Benin. Generally shallow, their water level varying with the tide, and their sandy shores fashioned by the waves into a multitude of small recurved spits, the lagoons are linked not only to the open sea but also to one another by small secondary channels which lead eastwards as far as Lagos. These channels are subject to slow infilling where their morphological evolution is not affected by external factors. The mud flats, covered at high tide, complete the coastal scene.

The chief dynamic factors affecting the morphological evolution of the coast of Dahomey are the tides, the surf and the offshore bar; these, inevitably, were fundamental problems involved in the selection of a

1 P. Pélissier, ' *Les pays du Bas-Ouémé* ', Travaux du Département de Géographie de Dakar, *no. 10 (1963)*.

suitable site for a new port. Hydrological investigations made between 1951 and 1961 showed that the tidal range of Cotonou is relatively slight, varying between a minimum of 1·2 m (3·9 ft) and maximum of 1·9 m. (6·2 ft). Hubert, however, recorded as early as 1905 that waves reaching a height of 4 m. (13 ft) above low-tide level caused great damage at Grand Popo.[1] More recent measurements of the swell, taken at the wharves at Cotonou and Lomé, indicate a variation between 1·5 m. (4·9 ft) and 3·4 m. (16·1 ft); and strongest swell occurs between June and September, with an average period of 11 to 14 seconds. The predominant direction is SSW.– SW. (52 per cent), with 36 per cent S.–SSW. and 12 per cent SE.–S. This swell originates in the South Atlantic and the Southern Ocean; Guilcher has pointed out that the swell is too strong to be of local origin in a tropical zone not characterised by typhoons or strong trade winds.[2] This hypothesis is supported by the fact that the swell persists in calm weather, by the diminution of the swell in the southern summer, and by the increase in the swell southwards along the African coast. On the Benin coast the swell is strongest between Fresco (Ivory Coast) and Axim (Ghana) and between the Volta and Niger deltas. The oblique orientation of the swell increases with proximity to the coast. One of the most important effects of the swell is the movement of sand along the coast in a west–east direction between Tabou and Port Bouet and between Cape Three Points and Lagos, although between Grand Bassam and Cape Three Points the direction is more variable. Beach and submarine profiles measured at Cotonou in connection with the problem of sand movement[3] have revealed the existence of a submarine trough situated between 40 and 110 m. (130 and 360 ft) from the shore. The swell breaks on either side of the trough, which occasionally disappears and re-forms elsewhere so that profiles at any given point may vary considerably during the year. On the landward side of the surf barrier the submarine slope becomes very gentle, and out to sea the currents are not strong because of the narrowness of the continental shelf. The problem of sand accumulation is directly linked with the phenomena of the submarine bar and the surf zone. The breaking of the surf along the line of the submarine bar places sand in suspension and initiates a complex easterly drift; in the upper levels of the water sand

1 P. J. Hubert, Mission Scientifique au Dahomey (*Paris, 1908*), *and* 'La barre au Dahomey', Annales de Géographie, *vol. 17, no. 92 (1908) 97–104.*
2 A. Guilcher, '*La région côtière du Bas-Dahomey occidental*', Bulletin de l'I.F.A.N., *série B, vol. 21, nos. 3–4 (Dakar, 1959), 355–425*
3 R. Garabiol, '*Les problèmes posés par la construction d'un port sur plage de sable. Étude du port de Cotonou, sur la côte du Benin*', Annales des Ponts et Chaussées (*Mai-Juin 1961) pp. 374–407.*

moves obliquely forwards in the direction of the swell, and in the lower levels travels back towards the open sea. In addition, a much weaker local swell is set up which generally breaks only on the beach, but which produces movement of sand along the beach in a zigzag manner. Although it has been estimated that $1\frac{1}{2}$ million cu. m. (53 million cu. ft) of sand is moved along the coast at Cotonou each year, the problem of sand accumulation was clearly one of the most important involved in the new port scheme.

Several elements in the coastal environment of Cotonou figure in the accounts of early travellers. According to one writer,[1] the settlement at Cotonou was at a distance of 350 m. (380 yds) from the sea in 1881 but only 50 m. (53 yds) in 1886. These figures may seem exaggerated, since a similar degree of sand removal is not observable today, but they nevertheless indicate the importance of mass sand transportation along the coast. Earlier, in 1875, Abbé Laffite had emphasised the difficulties of landing along the coast of Dahomey as a result of the sand banks which 'courent le long de cette côte' and of the 'barre dangereuse en tout temps'.[2] The Portuguese, who landed on the Slave Coast near Ouidah in 1580, were the first Europeans to brave these difficulties, but as a result of the inhospitable nature of the coast itself and the absence of gold, spices or ivory for trade purposes (slaves were not a major item of Portuguese trade at this period) Portuguese interest was directed elsewhere and only a few traders were left behind. The coast of Dahomey did not attract close European interest until the mid-seventeenth century, when Dutch, English and French traders were active in the area. French interests date from 1679, when Elbée's mission sent by Colbert arrived. The main cause of the growing interest in this stretch of the West African coast from this period onwards was of course the growing labour needs of Brazil, particularly after the discovery of mineral resources there.

Evolution of the port of Cotonou, 1893–1965

Before the building of the pier at Cotonou the unloading and loading of cargo on the Dahomey coast took place at the open roadsteads of Grand Popo, Ouidah, Djenkin (the modern Godomè Plage), Ekpé, (Semé) and, from the early nineteenth century, Agoué. Boatmen recruited from many points along the Guinea coast, especially from the Cape Coast (Ghana) area, crossed the bar with the aid of enormous surf boats, often throwing

---

1 *Abbé Laffite*, Le Dahomey: histoire, géographie, mœurs et coutumes, commerce, industrie. Expéditions françaises, 1891–94 (*Paris, 1895*).
2 *Abbé Laffite*, Le Dahomey, souvenir de voyage et mission (*Tours, 1875*).

themselves into the water in order to transfer goods and passengers from ship to shore and vice versa. These ports, described in detail by William Bosman[1] and Olfert Dapper,[2] were actively involved in the triangular trade of the Atlantic between France, West Africa and America, and the routes and cargoes of some of the ships concerned are recorded in detailed log books. For example, on 8 December 1772 the *Dahomet* left La Rochelle, arriving at Ouidah on 2 March 1773; from that date until 25 May she unloaded her cargo of textiles – cotton, silk, flax, velours, etc. – iron bars, cowries, pipes, hats, guns, spirits, knives and pearls. The ship then took on a cargo of 424 slaves later sold in Santo Domingo, whence a cargo of sugar, rum, tobacco and coffee was taken back to France.

After 1851 French commercial and political interests overseas became deeper and more interwined, and led in Dahomey to treaties of friendship and to the establishment in 1882 of a protectorate. In July 1889 a bi-monthly postal service was introduced between France and Dahomey, operated by the Compagnie des Chargeurs Réunis and the Compagnie Frayssinet. From August 1899 the service became monthly, and on the 11th of each month vessels such as the *Ville de Maranhao*, the *Ville de Maceio* and the *Ville de Pernambuco* left Le Havre bound for West Africa, calling at Bordeaux, Tenerife, Dakar, Conakry, Grand Bassam, 'Kotonou', and thence continuing to Libreville, Cap Lopez, Loango and Matadi.

The original pier at Cotonou was built in 1891–3 chiefly to receive materials destined for General Dodd's expeditionary force, designed to reassert French control over the Abomey kingdom. This pier had an open metallic framework and, like others along the West African coast, projected out to sea beyond the bar, but because of its open structure it did not form a barrier to lateral sand movement. A second wharf was built to replace the first in 1910, and in 1926–8 a landing platform 164 m. (538 ft) long was added, equipped with steam cranes. It was not until 1951–2 that the 1910 pier, almost at the end of its useful life, was extensively modernised. In 1954 the landing platform was lengthened and strengthened, and equipped with modern electric cranes; for the final decade of its operation the wharf was a well-equipped structure, with a pier 221 m. (725 ft) long and a landing platform 179 m. (590 ft) long, with six parallel railway lines and thirteen cranes.

The various improvements made to the wharf did not, however, greatly change the conditions under which cargo and passenger move-

1 *W. Bosman*, A New and Accurate Description of the Coast of Guinea (*London*, *1705*).
2 *d'O. Dapper*, Description de l'Afrique (*Amsterdam*, *1896*).

ments took place. For more than half a century both goods and passengers were landed by means of lighters, into which they were lowered over the ships' sides. Every traveller arriving by sea recalled with some feeling the descent from the ship and the subsequent crossing of the bar. Damage to goods handled in this way was estimated by the Chamber of Commerce to amount to between 2·50 and 2·75 per cent of the value of all goods handled. In spite of improvements in facilities at the wharf, vessels frequently had to wait several days before being worked; in 1954 waiting days totalled 556, or 37 per cent of the total number of ships' working and waiting days. Surcharges ranging from 15 to 100 per cent were frequently imposed by the shipping conferences in order to compensate for losses resulting from these prolonged delays.

Except during the war years traffic increased annually: from 95,000 tons in 1935, the total fell to 15,000 in 1939 and rose only to 65,000 by 1945; but by 1950 a figure of 146,000 tons had been reached (see Table 9.1), and by 1963 the total cargo tonnage handled reached 263,000. By this time the pier had effectively become incapable of handling a traffic flow of such dimensions, even though theoretically its capacity was estimated at 300,000 tons. From 1955 some traffic was diverted to Lagos – 53,000 tons in 1955, 70,000 tons in 1959 – but even so the practical limit had clearly been overreached. The building of a wharf at Grand Popo was considered as a possible solution, but abandoned as likely to be ineffective. From 1951 at least the inability of the pier at Cotonou to meet the requirements of the continually increasing traffic made itself felt, and a systematic propaganda campaign was organised in favour of its replacement by a deep-water port.

The new port: technical and locational problems

The recognition of the need for a new port on the Togo–Dahomey coast immediately introduced technical and political controversy. Studies made in 1952 by M. Pellenard-Considère[1] and by the Comité Technique des Travaux Publics de la France d'Outre-Mer in 1953 examined a variety of possibilities. Initially, the existing pier suggested the development of a lighterage port at the mouth of the Cotonou lagoon, a solution which would have continued the system of vessels anchoring in the open roadstead and would in fact have solved few of the port's problems. The difficulty of maintaining a minimum depth of 3 m. (9·8 ft) in the entrance channel was a further argument against this proposal. When attention

1 R. *Pellenard-Considère*, Rapport d'Études sur le port de Cotonou (*Bureau Central d'Études des Equipments Outre-Mer, 1952*).

turned to the question of a deep-water port, three solutions were discussed, all of which took account of the problem of sand accumulation and attempted to reduce as far as possible the necessity for expensive dredging in view of the quantities involved and the lack of suitable equipment. Models of the three possible solutions were developed in the laboratories of the Société Grenobloise d'Études Agricoles et Hydrauliques (SOGREAH).

The first model was based upon the principle of accepting sand accumulation to the west behind a breakwater, the presence of which would stimulate sand erosion towards the east. Tests showed that, assuming an annual movement of $1\frac{1}{2}$ million cu. m. (53 million cu. ft) of sand, and an angle of swell of 13 degrees, all easterly movement of sand would be arrested for a period of twenty-five years by a 1,500-m. (1,577-yd) breakwater. Thereafter sand movement past the harbour would progressively increase from 20 per cent of the normal volume in the thirty-sixth year to 50 per cent in the forty-fifth year and to 75 per cent in the sixty-fifth year. A fairly close correspondence between sand erosion and sand accumulation was observed on the model, except in the immediate vicinity of the port installations. A second model introduced the idea of artificial sand removal, with dredging on an extensive scale west of the port and dumping to the east, thus avoiding the problems associated with sand accumulation inherent in the first model but introducing a laborious and expensive mechanical alternative to a natural process. The third proposition focused upon the idea of an artificial island port sited out to sea beyond the surf barrier and linked to the mainland by a pier. This proposal was rejected, however, on the grounds that it was not only the most expensive of the three but also because it provided the most limited amount of working space. The first proposal was eventually selected in spite of its long-term disadvantages, because it avoided the dredging problem and because the area of sand accumulation to the west of the main breakwater could eventually be used for port extensions and urban expansion.

Political controversy surrounded the decision to develop a new port because physical conditions vary only slightly along the coast of the Bight of Benin and no overriding technical arguments could therefore be adduced in favour of a specific port site. Arguments developed at two levels: the inter-territorial level between the then dependencies of Dahomey and Togo, and the regional level within Dahomey itself. Many factors favoured the development of a single new port to serve both Dahomey and Togo, perhaps at Grand Popo or Agoué. The existing railways serving Cotonou could have been prolonged westwards along the coast, the two political and economic capitals of Cotonou and Lomé would have been roughly equidistant from such a scheme, and phosphate deposits close to the Togolese coast at Anécho, east of Tsévié, would have been conveniently served. Opposing arguments were raised by economic interests in Cotonou itself, a town equipped with physical installations costly to transfer elsewhere; the construction of a new town at Grand Popo and the modification of existing economic systems would have entailed considerable additional expenditure. Within Dahomey, political interests and electoral issues supported various alternative sites – Semé, Avrékété and Ouidah as well as Cotonou itself. Eventually, on 10 September 1957, the Territorial Assembly finally selected Cotonou and directed that the new deep-water port should be constructed there, immediately to the west of the old pier. Work began early in 1960 and was completed in 1965 by a group of French enterprises under the control of the Bureau Central d'Études pour les Équipements d'Outre-Mer (B.C.E.O.M.).

Facilities and equipment of the new port

The new deep-water port of Cotonou (Fig. 9.2; Plate 10) is equipped with four principal berths, each 155 m. (508 ft) long, aligned parallel to the coast and forming a continuous 660-m. (2,165-ft) quay. Five vessels of moderate length can be accommodated, although four is usual. The berths are protected from swell and littoral drift by the 1,210-m. (3,970-ft) main breakwater which is built on rubble foundations with tetrapod armour and which curves eastwards at its seaward end. An eastern breakwater, 770 m. (2,525 ft) long, is also based on rubble at its shoreward end but at its seaward end is built of sheet-steel piling and concrete. At the tip of this breakwater are two oil berths, one of which is intended to serve as a shelter berth in bad weather. Two other berths are situated on this break-water, one used for bulk palm-oil shipments and the other by occasional small trawlers. This fishing berth forms the nucleus of the proposed fishing port, which is to have a special 90-m. (295-ft) quay, a storage area of about 20,000 sq. m. (216,000 sq. ft), and a zone where fishing craft can be beached. There will also be an open fresh-fish market and a cold store. A 240-m. (262-yd) railway branch line will serve the fishing port and will permit Dahomey, like its neighbours Ghana and Nigeria, to develop a well-organised fishing industry. Each of the four main berths is equipped with a transit shed with a floor area of 3,500 sq. m. (37,560 sq. ft), and behind the sheds there is considerable open stacking space. The quayside

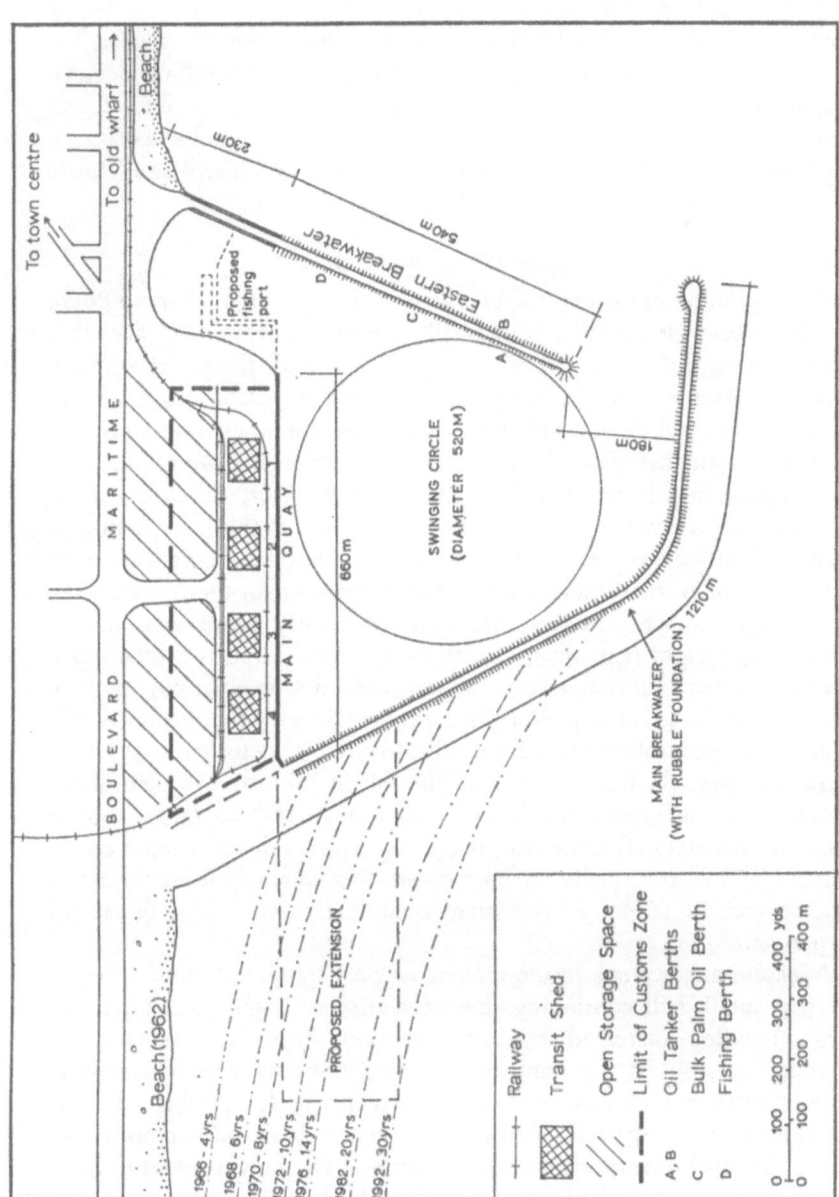

Fig. 9.2 The port of Cotonou: layout and facilities

is served by a single-rail track, and the landward side of the shed has double tracks. All loading and unloading of cargo is effected by ships' gear, but the port is well equipped with small-scale mechanical handling devices. The theoretical cargo-handling capacity of the new port is 1 million tons a year, and with an annual total traffic flow averaging around 450,000 tons at present it seems that Cotonou provides facilities which will be adequate to meet Dahomey's requirements for some years to come.

The growth of port traffic

At the beginning of the Second World War the surf ports of Grand Popo, Ouidah, Porto Novo and Agoué finally ceased to operate, and since that time virtually the whole of Dahomey's external trade has passed through Cotonou. Records kept by the port authorities, the Dahomey Chamber of Commerce and the Organisation Commune Dahomey–Niger provide the basis for an analysis of the growth of port traffic at Cotonou. Several major phases may be distinguished. The period 1935–49 was characterised by considerable overall growth and by a very favourable export–import balance. Within this period, the pre-war years 1935–9 showed the most rapid increase in total traffic handled, the annual rate for these years being on average 12 per cent. Exports were frequently more than double imports by weight (e.g. 1935, exports 65,630 tons, imports 29,020 tons). Taking the 1935 figures as 100, exports reached a level of 167 in 1939, compared with 142 for imports in the same year. During the Second World War the general traffic level fell considerably, to 99,342 tons in 1941 and to 66,400 in 1945, chiefly as a result of the fall in imports from European countries. Total imports fell from 85,443 tons in 1941 to 13,975 tons in 1945. Immediately after the war, the years 1946–9 showed a rapid rise in port traffic which virtually attained its pre-war level within three years – 147,260 tons in 1949 – representing an annual average growth rate of 32 per cent.

A second major phase in the growth of port traffic at Cotonou began in 1949 and is still continuing. Between 1949 and 1967 the volume of cargo handled increased by 200 per cent, from 149,000 tons to 441,000 tons, but at the same time the previously favourable export–import balance was sharply reversed. Except in 1950, when climatic conditions were especially favourable, exports have been consistently below the level of imports during this period. The present unsatisfactory economic situation in Dahomey thus dates back at least twenty years as far as trade is concerned.

A breakdown of export commodity figures reveals that palm-oil

Table 9.1

Traffic at the port of Cotonou, 1948–67
('000 tons)

Year	IMPORTS			EXPORTS					Total Traffic
	Construction materials	Foodstuffs	Total imports	Palm kernels	Palm oil	Ground-nuts	Cotton	Total exports	
1948	7·8	7·3	70·5	37·6	11·4	—	—	78·2	148·7
1949	27·0	10·6	79·8	44·4	11·4	3·1	4·0	70·0	149·8
1950	21·8	12·4	63·5	46·5	12·3	8·8	1·7	79·8	143·3
1951	27·4	17·5	65·7	34·2	14·6	3·2	1·5	63·5	129·2
1952	33·2	10·2	87·6	37·9	9·0	2·3	1·0	65·7	153·3
1953	35·7	18·2	91·1	47·0	16·4	4·3	0·7	87·6	178·7
1954	43·0	27·9	110·1	48·2	17·3	4·7	1·1	96·1	206·2
1955	46·3	25·3	121·4	50·9	17·1	10·5	0·7	110·1	231·5
1956	56·8	28·6	110·8	54·0	19·8	15·5	1·0	121·4	232·2
1957	70·0	23·0	134·1	44·6	13·5	12·7	0·8	110·8	244·9
1958	57·0	28·6	105·3	60·7	18·6	13·4	1·3	134·1	239·4
1959	63·3	33·1	125·6	50·4	12·8	1·6	1·3	105·3	230·9
1960	90·4	38·8	115·0	61·3	15·4	12·9	2·2	104·6	219·6
1961	59·7	30·2	157·0	41·5	11·5	12·5	1·3	93·4	250·4
1962	56·9	33·7	159·5	43·9	9·3	4·3	0·6	75·0	234·5
1963	68·7	47·3	180·0	50·5	9·2	6·5	1·4	83·2	263·2
1964	57·3	36·1	180·3	56·1	12·7	3·9	1·1	89·7	270·0
1965	85·5	37·1	227·5	14·3	13·2	15·6	1·2	93·9	321·4
1966	86·5	55·2	265·8	8·4	12·3	29·5	2·7	103·9	369·7
1967	85·9	53·2	289·7	4·8	7·6	82·49	3·5	157·2	446·9

Source: *Direction du Port de Cotonou.*

produce, together with other oils and oilseeds (especially groundnuts), dominate the traffic of the port, constituting 80 per cent of the total volume of exports in 1949, 88 per cent in 1955 and 92 per cent in 1965. Oil-palm production was initially encouraged by the Dahomeyan ruler Ghezo (1818–58) who replaced the export of slaves by the export of palm produce. By 1890 Dahomey already exported 5,225 tons of palm-oil and 14,653 tons of palm kernels; by 1938 these figures had risen to 9,959 tons and 38,572 tons respectively. The peak export figures achieved in 1936, when 19,500 tons of palm-oil and 74,000 tons of palm kernels were exported, has not yet been exceeded despite active government support for the palm industry. Since palm produce forms such a high proportion of the total volume of exports, the changing pattern of palm exports is inevitably closely related to that of the overall export situation, with generally rapid growth leading up to the 1936 peak, annual fluctuations depending mainly on rainfall variations. The war years were inevitably low years for palm exports – 4,000 tons in 1943, 2,500 in 1946 – but the immediate post-war increase is noticeable. Although exports of palm produce have gradually increased during the 1949–67 period, the pre-1939 totals have not been exceeded. The changing balance of private and estate production is another interesting feature of the palm industry. In the pre-war years, and until 1952, all exports of palm produce came from small local peasant farmers, but after 1952 estate production gradually increased in importance from 1,900 tons (1952) to 11,000 tons (1960), whilst at the same time the export production of peasant farmers decreased from 11,000 tons in 1948 to 4,000 tons in 1960. These changes are linked with rapid population growth, involving a fast-increasing domestic palm-oil consumption, and with the development of several large palm-oil mills.

Groundnut exports (mainly from Niger), which stood at 3,972 tons in 1936, reached a record figure of 82,475 tons in 1967. Apart from these principal items, other commodities entering significantly into the export trade of Cotonou have shown a static or declining tendency in recent years; these are mainly coffee, cotton, castor oil, shea-nuts and kapok. Shea-nuts, for example, have declined from 13,200 tons in 1948 to 900 tons in 1960 and amounted to only 2,445 in 1967. Maize exports, important in 1936 (33,000 tons), ceased completely during the war period. It was only in 1967 that the total export tonnage exceeded the record established in 1936 (133,000 tons).

Building materials, especially cement, are the most important items imported at Cotonou, forming 52 per cent of the total volume of cargo imported. Other major sectors of the import traffic include foodstuffs

(31 per cent in 1967), textiles (especially printed fabrics), and household goods. Petroleum products have been imported at Cotonou since July 1963, well before the opening of the new port. Previously, from 1952, bulk oils were unloaded at Lagos and redistributed thence to Dahomey, Niger, Togo and eastern Mali. In 1965 the AGIP company opened an oil depot in Togo from which products are distributed in Dahomey, and reciprocal arrangements exist for companies based in Dahomey, so that road tankers travel in both directions.

For many years traffic to and from the Niger Republic has been important at Cotonou, and has in fact increased more rapidly than Dahomey traffic in recent years. This is partly due to the introduction of the 'Operation Hirondelle' system, under which a seasonal north–south road service is provided between the southwestern part of the Niger Republic and the Dahomey railhead at Parakou. In 1949 Niger traffic formed only 9 per cent of the total volume of cargo traffic handled at the Cotonou wharf, but by 1966 the proportion had increased to 35 per cent. The main items involved are exports of groundnuts and cotton and imported manufactured goods. Most traffic to and from the eastern and central areas of Niger, however, passes through Lagos via Zinder or Maradi and the Nigerian railway system; and western Niger is also served by the port of Abidjan, both by road and rail (via Ouagadougou) and directly by road (Abidjan–Niamey, 1,730 km. [1,080 miles]). The competition of Lagos and Abidjan may have a serious effect upon the growth of traffic at Cotonou unless freight rates are amended in favour of the new port.

Passenger traffic at Cotonou has shown a marked decline in recent years (Table 9.2). This is partly due to the fact that since political independence the number of Dahomeyan officials working outside the country in both the public and private sectors has been greatly reduced by the policies of nationalisation of personnel adopted in the former French states of West and Equatorial Africa. In addition, the repatriation of French military and civilian personnel formerly stationed in Dahomey has drastically reduced the number of French citizens travelling by sea to France for their annual leave. Because of these factors, the number of passenger vessels calling at Cotonou has dropped. Another characteristic feature of the structure of port traffic at Cotonou is the dominant position held by the franc zone, especially by France itself, in Dahomey's external trade. Although the proportion of trade with countries in the dollar and sterling zones has increased since independence, the former metropolis still holds a predominant position (Table 9.3).

Table 9.2

Passenger traffic at Cotonou, 1959–67

Year	Passengers disembarked	Passengers embarked	Total
1959	5,429	6,100	11,529
1960	5,493	4,208	9,701
1961	6,144	4,996	11,140
1962	7,051	4,829	11,880
1963	4,851	5,004	9,855
1964	5,182	5,284	10,466
1965	4,178	5,008	9,186
1966	4,117	4,367	8,484
1967	3,806	4,116	7,922

Source: *Direction du Port de Cotonou.*

Table 9.3

Traffic of the port of Cotonou with France and with the franc zone as a percentage of the total traffic of the port, 1959–67

Year	Exports Franc zone	France	Imports Franc zone	France
1959	75·0	69·0	72·0	61·0
1960	84·3	67·0	77·7	58·6
1961	90·1	72·0	76·7	54·4
1962	84·6	70·4	75·2	58·4
1963	81·3	71·5	73·2	61·1
1964	79·8	75·0	69·3	57·7
1965	63·1	54·7	65·1	54·0
1966	73·0	43·0	63·0	51·0
1967	47·0	31·0	53·0	50·0

Source: *Direction du Port de Cotonou.*

The trade of Cotonou is thus characterised primarily by a serious import–export deficit, particularly noticeable in recent years. Janin[1] has suggested three main reasons for this commercial situation: population growth and urban expansion, and consequent increased demands for consumer goods; an increased market for building materials, especially in the context of the renewal of the port itself since 1960; and the reduction of agricultural surpluses, the obsolescence of the palmeries and the recent bad harvests which have caused the output of cotton, groundnuts and coffee to fall seriously. It may be argued, however, that population growth has only slightly affected imports, for the gross purchasing power does not necessarily increase directly in proportion to the number of people. Population increase, especially in urban areas, is certainly a major feature of the changing economic pattern in Dahomey, but the standard of living of peasant farmers appears to have declined during the period 1949–65 by between 15 and 25 per cent. This situation is clearly linked with declining agricultural output which, in spite of population growth, prevents increased consumption of both local products and imported goods. Additional factors in the situation include the excessive profits frequently made by the African middlemen in Dahomey and, over the past few years, the marked deterioration in exchange rates between developed and developing areas of the world. At Cotonou, the Direction des Affaires Économiques has estimated that during the period 1960–4 the ratio between the average value of 1 ton of imported goods and that of 1 ton of exported goods rose from 1:0·95 to 1:1·25, representing an increase of 25 per cent in the value of imports.

Declining exports of coffee and groundnuts arise in part from increasing concentration of production elsewhere, particularly in Senegal and Ivory Coast where more favourable natural conditions permit a greater degree of specialisation of crop production. The oil-palm plantations of Dahomey are of great age, and government-sponsored attempts at regeneration have not always been successful. Price stabilisation and measures intended to limit the purchase of palm produce from countries outside the former French Overseas Territories have only been introduced at a rather late stage.

A further factor affecting the unfavourable balance of trade situation in Dahomey is the occurrence of illegal traffic movements across the boundaries with Nigeria and Togo. The value of this illegal trade, estimated by the Cotonou Chamber of Commerce to be more than

1 B. Janin, 'Le nouveau port de Cotonou', Revue de géographie alpine, *vol 52, no. 4 (1964) 701–12.*

£3 million per annum, inevitably reduces considerably the traffic of the port. The main commodities involved in this international smuggling are textile goods, cigarettes and matches, enamel ware, bicycles, sewing machines, radios, tinned foods, wines and spirits and perfumes. This traffic is facilitated by the ready permeability of the boundaries between Dahomey and Nigeria and Togo, and between Togo and Ghana, and also by the often considerable disparity between customs tariffs in Dahomey and its neighbours. In spite of government efforts to reduce this disparity by reducing certain taxes on cigarettes, spirits and textiles, the problem of illegal trading has by no means been solved, and customs authorities have at times been prevented by political pressures from carrying out their functions adequately.

Looking towards the future it seems clear that the facilities provided by the new port of Cotonou will not be fully utilised until the economy of Dahomey is much further advanced. Efforts have been made in recent

Table 9.4

Port of Cotonou: projections for Dahomey and Niger traffic, 1965–85
('000 tons)

Year	Dahomey traffic	Niger traffic	Total
1965	230	70	300
1970	270	110	380
1975	360	140	500
1980	480	170	650
1985	630	200	830

Source: R. *Condomines and H. Leroux*, L'avenir du port Cotonou (*Paris, 1961*).

years to revitalise the oil-palm industry, whose products are likely to remain the chief exports of the port for some time. Several large co-operative estates of selected trees have been established with government aid under the control of the Société National du Développement Rural (SONADER) at Houin (Mono département, 4,000 hectares [15·5 sq. miles]), Hinvi (Atlantique département, 6,000 hectares [23 sq. miles]), and Agouvi (Ouémé département, 6,000 hectares [23 sq. miles]). Oil mills have been built to serve these plantations, and it is estimated that the output of palm produce can be doubled or even trebled in the relatively near future. In addition, encouraging results have been achieved from experimental cotton and groundnut plantings in other parts of the country. In international terms it seems likely that traffic to and from Niger will assume an increasingly large share in the trade of Cotonou. Projections for Dahomey and Niger traffic at Cotonou (including petroleum products) are shown in Table 9.4. The question of using Cotonou as an overflow port for Lagos has been frequently raised, and Cotonou could undoubtedly provide an effective service for western Nigeria as well as for the Niger Republic. It is unlikely that any question of competition would arise if this solution to the problem of recurrent congestion at Lagos were adopted, because of the clear distinction between the traditional hinterlands of each port.

Some effects of the development of the new port

The opening of a new port clearly has complex repercussions in a variety of spheres. In physical terms, the permanent opening of the Cotonou channel has modified water movements and increased marine influences in Lake Nokoué, where the shores are now occasionally subject to erosion. The annual floods of the Ouémé river, which reach the sea through the Cotonou channel, have been regularised to some extent, and the biological environment and the fishing economy of the lake and lagoon areas have been substantially modified. Recent studies made by the Centre Technique Forestier Tropical have discovered a marked reduction in the growth rates of certain fish species such as *Tilapia* owing to changes in the occurrence of certain chemical elements in the water. These changes have, however, encouraged the growth of molluscs and shellfish. More broadly, the entire geographical environment of the lagoon zone is undergoing modification as a result of the permanent opening of the channel. Fish catches are diminishing, and some traditional forms of fishing have been abandoned. Unemployment has become a serious problem amongst the coastal populations around Lake Nokoué and the Porto Novo lagoon. Technical solutions to the problem of the declining fishing industry, and social solutions to the problems of unemployment – resettlement and redeployment – are under active scrutiny. An additional social problem resulting from the opening of the new port concerns the re-employment of the various employees at the old wharf; these include not only a variety of officials and agents but also some 250 lightermen, many of whom came originally from Ouidah and Grand Popo. A commission set up with the object of retraining these workers as fishermen within the national co-operative movement has so far been a complete failure, mainly because previous experience has led the men to have little faith in the co-operative system.

The opening of the new port has inevitably affected the town of Cotonou in a variety of ways, but in overall terms has provided a marked stimulus to urban growth. Together with other urban areas in Dahomey, Cotonou has naturally played a large part in the growth of imports at the new port, and the increased facilities for importing building materials of all kinds have acted as a direct stimulus to urban growth. The rapid growth of Cotonou in terms of population and in physical terms is clearly linked very closely, although by no means solely, with the development of the new port. In recent years successive areas of Cotonou have been turned virtually into builders' yards. Road communications have also been improved, especially in and around the port area and within the commercial zone of the town. The chief outstanding problem is that of the Cotonou Bridge, which provides the only link across Lake Nokoué. More than 2,000 vehicles, and thousands of cyclists and pedestrians, use the bridge every day, and congestion at this point is beginning seriously to affect the working of the port, particularly since the bridge links the port area and the industrial zone of Akpakpa. A new bridge 2 km. (1·2 miles) to the north is proposed, when funds become available. Another transport problem concerns the railway, which at present bisects the town; a new branch now terminates at the deep-water port. At the northern end of the line there is strong support for an extension of the Parakou line at least as far as Dosso in view of the increasing importance of traffic to and from Niger.

Certain important changes in the operation and administration of the port have been introduced with the opening of the deep-water facilities. At the old wharf, cargo handling was effected by the Organisation Commune Dahomey–Niger (O.C.D.N.); in the new port, movements are controlled by the Groupement des Manutentionnaires du Dahomey (G.M.A.D.A.), a consortium of three shipping agencies. Port operation thus appears to have passed out of direct state control into the hands of powerful commercial interests. An average reduction of 30 per cent has occurred in port charges in comparison with the old wharf; on certain goods such as cement and food products import taxes have been lowered by as much as 50 per cent. However, these reductions have not been passed on to the consumer, partly because traders' profits have increased and partly because the government has imposed a variety of other supplementary taxes.

Conclusions

In technical terms the port of Cotonou is clearly a successful development, in so far as effective use has been made of an unfavourable site and the problem of littoral sand movement and accumulation has been solved. At no other part of the coast was there a strictly comparable problem. At Takoradi, the oldest of the artificial ports, the volume of transported sand is much less. At Abidjan and Lagos, although the sand movement is greater the ports have been located on the shores of sheltered lagoons and not on the open coast. The struggle against sand accumulation is a delicate problem facing many ports in different parts of the world; outside West Africa, examples include Safi (Morocco) and Mahendy (Tunisia). In Chile, the development of the port of Constitución has been severely checked by sand accumulation, which has lengthened a coastal spit by almost 1 km. (0·62 mile) in less than ten years. At Cotonou the problem seems to have been solved at least temporarily, and the port will be able to double its present capacity when a basin is excavated in the sand which accumulates to the west of the main breakwater. The principles of port layout followed at Cotonou are not unlike those developed at Tema (Ghana), and a similar scheme has recently been completed at Lomé (Togo).

A variety of economic problems has been raised by the development of the new port, for the economic function of a port can only be fulfilled in the wider context of the changing agricultural, industrial and social environment of the tributary area. Present-day Dahomey has not yet reached the take-off point in terms of modern economic growth, and in this context there is a sharp contrast with certain other West African states. Undoubtedly, both natural and human potentialities exist for development; Dahomey, although small, is a coastal state enjoying a wide range of climatic and agricultural environments as do the other coastal states of West Africa. In the south at least, Dahomey possesses long-standing skills and cultural attributes of great value; yet the country is generally regarded as poor. Political instability is frequently said to have prevented large-scale investment, but it is sometimes forgotten that potential investors are primarily influenced by more fundamental economic considerations. Since independence, Dahomey has never made a precise choice as far as her development goals are concerned. Various development plans have been drawn up, but none has ever been effectively worked through because the initial range of economic activity has always been too restricted. Since the days of the slave trade, Dahomey has had an essentially export-based economy; as in most other states of tropical Africa, little capital is available for development; plans are always drawn up on the assumption that funds will eventually become available, but to date many hoped-for subventions have failed to materialise.

In the commercial sphere, so closely linked with the activities of the port, a fundamental reorganisation is necessary. The existing commercial structure in Dahomey is anachronistic, and the anarchic situation in the local markets causes the prices of many consumer goods to vary by as much as 100 per cent from one side of the street to the other! Problems of price control are currently as insoluble as problems of inter-state smuggling in Dahomey. The long-term success of the Cotonou port project depends in part upon the successful solution of these difficulties, as well as upon the economic and political relations with Niger. Political tensions have in the past adversely affected the traffic of the port of Cotonou, and may well do so again; but for the present friendly relations appear to be well established, and it is to be hoped that these will continue and will permit some degree of inter-state regional economic co-operation utilising the port of Cotonou as a common base. There is a close parallel with the neighbouring port of Lomé, now undergoing a similar transformation to that seen at Cotonou; Lomé will continue to depend to a considerable extent upon traffic to and from Upper Volta when the new facilities are opened. In broad terms, however, the two ports of Cotonou and Lomé are likely to experience only modest future growth in comparison with their larger neighbours on the Gulf of Benin – Abidjan, Tema and Lagos.[1]

1 *For further information on Cotonou, see J. d'Almeida, 'Notes sur les données économiques au Dahomey', cyclostyled report (1966); P. Chauleur, 'À propos du port de Cotonou', Industries et Travaux d'Outre-Mer, vol. 60 (1958) 649–50; M. Giraud, 'Le futur port de Cotonou', Industries et Travaux d'Outre-Mer, vol. 55 (1958) 324–7; Government of Dahomey, Rapports Annuels du Service du développement rural (1963 ff.); J. Pososki, Étude des transports urbains au Dahomey, U.N. report (1965); M. Thomann, 'Les transports maritimes au Dahomey', Cinquentennaire de la Chambre de Commerce de Cotonou (1958).*

10. Patterns and Problems of Seaport Evolution in Nigeria

Babafemi Ogundana

Mr Babafemi Ogundana, M.Sc., is Lecturer in Economic Geography with special reference to transportation in the University of Ife, Nigeria.

This chapter is divided into four parts. The first part develops some concepts and a model of port evolution; the second examines the process of port evolution in Nigeria; the third provides a summary of the factors which interplay to structure the pattern of port development in Nigeria; whilst the fourth part briefly considers the planning and applied problems which result from changing patterns of port development.[1]

Element in a model of port evolution

Seaports are dynamic phenomena, changing not only in their morphology but in their functions and status. Changing port significance may be considered first in an absolute and secondly in a relative sense. Changing absolute significance concerns the varying influence of the particular port by itself over time; in relative terms one port is viewed in comparison with other ports. As any single port is not an isolated unique feature, its character is best appreciated in a comparative treatment in relation to other ports, especially in its neighbourhood.[2] A set of ports which are *related* can be referred to collectively as a *port complex*. Port relatedness rests on functional association and interdependence, and can be viewed from a maritime or from a continental aspect. On the seaward side, ports served by the same shipping services and thus linked to common forelands are related. On the land side, ports that can serve as alternative outlets to a part or all of a defined unit area are, within that area, related. The relationship of such ports may be competitive or complementary. In a competitive

---

1 *I am grateful to Professor James Bird who encouraged me to write the first version of this paper, and for his advice and that of Professor R. J. Harrison Church on the substantive work from which the paper is derived. I am grateful too to Professor A. L. Mabogunje and Dr G. J. A. Ojo for beneficial academic stimulation over many years. I must also thank the editors of this volume for their co-operation.*

2 N. Manfred Shaffer, The Competitive Position of the Port of Durban (*Evanston, Ill., 1965*).

framework, the related ports vie for traffic in their common hinterland, and developments at any one port affect the fortunes of the other ports in the complex. When they complement one another, particular ports may be devoted to the handling of particular commodities from the common hinterland.

Ports may be associated, and may form complexes at varying scales, local, national or extra-national. The western Niger Delta ports of Koko, Sapele, Warri and Burutu are an example of a local complex. Seaports in Nigeria, as a whole, form a larger complex (Fig. 10.1). The general cargo

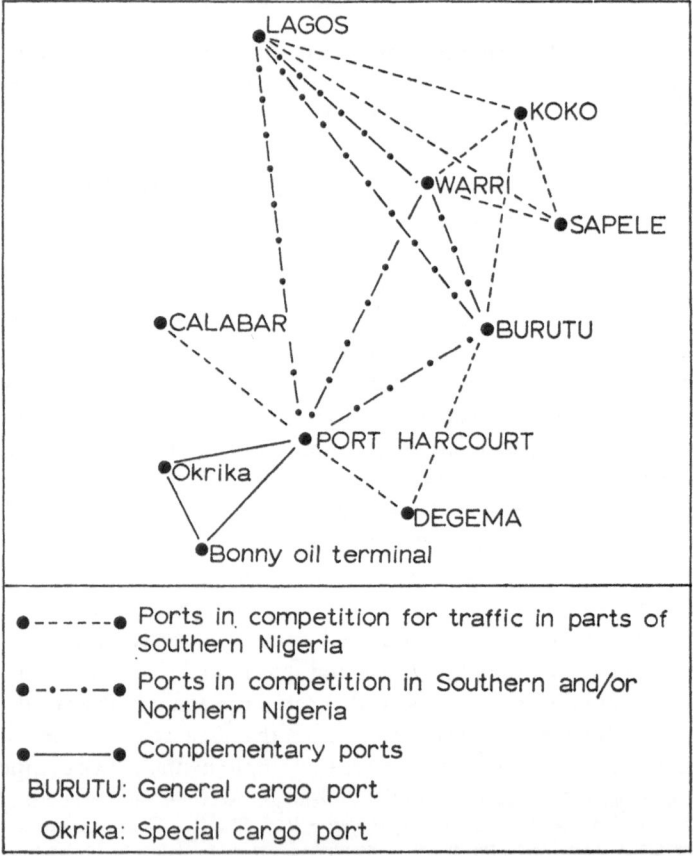

Fig. 10.1 Functional relationships in the Nigerian port complex

ports compete for traffic in southern and/or northern Nigeria. The two special cargo ports have come into being only since 1960. The Bonny oil terminal handles crude petroleum, and Okrika handles refined oil products. The two can be seen as complementary to each other, and to Port Harcourt, the 'parent' port in the area. The competitive relationships of Nigerian general cargo ports are not a recent phenomenon, but a feature of the pre-twentieth-century period also. On the supranational scale, major West African ports are associated by common shipping flows and are in competition, especially in respect of the traffic of the landlocked states.

At any point in time, the ports in a complex are of unequal significance and may thus be ordered into a hierarchy. The character and functions of the various ports in a complex change differentially over time, owing to the unequal impact upon them of factors influencing port evolution. Each port may be of either increasing or decreasing relative significance,[1] and ranking of the ports in the complex changes as the relative significance of the individual ports alters.

As the absolute and relative significance of the individual ports change, so also does the composite character of the complex taken as a whole. It is from the sequence of changes in the ordinal ranking of ports in a complex that one can gain an idea of the pattern of evolution of the complex taken as a unit. The composite pattern of a port complex, over time, may take one of two basic forms: *port concentration* or *port diffusion*. Port concentration implies that a few of the many ports in a complex are of disproportionate significance, and this structure emerges as a result of the increasing relative significance, over a period, of certain ports as compared to others which either gain more modestly or decline in absolute terms. Port concentration results initially in relative decline but may lead to an absolute decrease in the number of ports operating. Port diffusion occurs when higher-order ports decline, leading to the increasing significance of new or previously smaller ports, and involves an absolute or relative increase in the number of functioning ports. New ports may develop to serve new trade, or part of an existing trade, giving an absolute increase in the number of ports, or there may be a diversion of trade to formerly less important ports.

Each era of concentration or diffusion has its own port hierarchy, and leading ports may change from one period to another. A complex in which the hierarchy undergoes considerable change may be termed an *unstabilised port structure*. However, a complex may attain stability with one or more points attaining leadership (though with fluctuating relative significance) during successive periods of concentration and diffusion.

1 *See, for example,* G. G. Weigend, '*Bordeaux, an example of Changing Port Functions*', Geographical Review, *vol. 35 (1955) 217.*

Some of the leading ports in Europe and North America, for example, have maintained relative leadership in their regions during the last two centuries. Such persistent centres of significance may be termed points of *sustained port dominance*. When such persistent centres of significance emerge, a basic form of *spatial consolidation* of port significance is created. It is the idea of spatial consolidation or port concentration that has received most frequent mention in the literature, and an elucidation of the underlying process of alternating concentration and diffusion is rare. Sargent gave a general statement of this phenomenon when he observed that 'the tendency of modern sea traffic is to concentrate so that the number of effective ports grows less'.[1] A number of regional studies (for example Gould for Ghana, Spate for India, Stamp and Beaver for the United Kingdom) have also observed the feature of spatial consolidation of port functions.[2] Patton appears to have given a first indication of the underlying processes of concentration and diffusion in the structuring of port complexes. He observed that 'the secular trend towards the consolidation (better concentration) of American import and export trade in fewer and fewer ports has gradually been halted and then reversed, presenting opportunities for new ports and competitive problems for ports endeavouring to retain their position'.[3] The reversal of concentration and the relative pull-out of trade to new or previously declining ports initiates a pattern of port diffusion. In long-term perspective, the concentration or consolidation into fewer and fewer ports is not a continuous process, but falls into recognisable phases and is periodically interrupted to yield a cyclical evolutionary sequence in which a measure of concentration alternates with some form of diffusion. Accompanying this alternation are changes in the port hierarchy, and underlying both is the changing significance of individual ports. When a port assumes sustained dominance, alternating concentration and diffusion merely reflect an established inequality in the spatial disposition of port functions.

One important question in studying the structure of port complexes is that of measurement. How can one measure absolute and relative port significance, and how is one to derive an index of the composite structure of the whole complex? The ranking of ports depends on the criteria that are taken as indicating port significance. The relative merits of various

1 *A. J. Sargent*, Seaports and Hinterland (*London, 1938*) *p. 2*.
2 *O. H. K. Spate*, India and Pakistan (*London, 1957*) *p. 325; L. D. Stamp and S. H. Beaver*, The British Isles (*London, 1962*) *p. 653; Peter R. Gould*, The Development of the Transportation Pattern in Ghana (*Evanston, Ill., 1960*).
3 *D. J. Patton, 'Recent Literature on Ports'*, Annals of the Association of American Geographers, *vol. 47 (1957) 194*.

criteria have been reviewed elsewhere.[1] Just as the size of population could provide a first indication of the status of a settlement, volume of port traffic, measured in tonnage or value, could be used as an immediate though necessarily imperfect criterion in comparing port significance. Other elements in port status could subsequently be evaluated in relation to this first index. It is necessary to break down the total port traffic, not only into loaded and unloaded movements, but further into particular commodity flows.[2] If the volume of traffic is taken as an index of absolute significance, changes in the relative significance of ports in a complex can be measured by calculating, for different years, the percentage share of individual ports of the total trade for the group.[3] Having quantified changes in the absolute and relative senses in relation to individual ports, the next problem is to measure the composite structure of the complex.

As pointed out above, a first indication of a change in the structure of a complex is given by a change in the absolute number of ports, a decrease reflecting concentration and an increase indicating diffusion. It is, however, the relative share of trade by the different ports in the complex that can accurately reflect the structure of the complex. Hirschman's index of trade concentration can be adapted to measure port concentration or diffusion.[4] The index, I, is given as the square root of the sum of squares of the percentage shares of all markets (in this case ports) in the trade. A period when there is consistent decrease in the value of the index can be identified as one of diffusion, whilst a period when the values increase is one of concentration. Changes in the magnitude of the index can also indicate the intensity of concentration or the degree of diffusion.

The process of port evolution in Nigeria

The process of port evolution in Nigeria over the past four centuries exemplifies the ideas outlined above; evidence is provided of changing port significance, alternating concentration and diffusion, and spatial

1 F. W. Morgan, Ports and Harbours (*London, 1958*) *pp. 13–21; James Bird,* The Major Seaports of the United Kingdom (*London, 1963*) *p. 21; R. E. Carter,* 'A Comparative Analysis of United States Ports and their Traffic Characteristics', Economic Geography, *vol. 38 (1962) 162–75.*
2 G. G. Weigend, 'Some Elements in the Study of Port Geography', Geographical Review, *vol. 48 (1958) 185–200.*
3 C. Chen, 'The Port of Keelung', Tijdschrift voor Economische en Sociale Geografie, *vol. 48 (1957) 142–8.*
4 A. O. Hirschman, National Power and the Structure of Foreign Trade (*Berkeley and Los Angeles, 1945*) *pp. 157–62; cited by D. H. Jones, 'Exports and Imports in the Domestic Setting', in R. G. D. Allen and J. E. Erley (eds),* International Trade Statistics (*New York, 1953*) *p. 237.*

consolidation in port development. Six periods of alternating concentration and diffusion have emerged from an empirical study of changing port significance in Nigeria (Table 10.1).[1] During the first period European

Table 10.1

The pattern of seaport evolution in Nigeria

Phase	Time period	Composite form	Leading ports in hierarchy	Overall trend
I	1500–1670	Initial concentration	Gwata, Bonny, Old Warri	Unstabilised port structure
II	1670–1750	Diffusion	Old Calabar, New Calabar, Bonny, Brass	Unstabilised port structure
III	1750–1860	Concentration	Bonny, Lagos	Unstabilised port structure
IV	1860–1910	Diffusion	Lagos, Akassa, Old Calabar, Sapele, Warri, Degema	Consolidation with Lagos as point of sustained port dominance
V	1910–1950	Concentration	Lagos, Port Harcourt	Consolidation with Lagos as point of sustained port dominance
VI	1950–	Diffusion	Lagos, Port Harcourt, Bonny, Okrika	Consolidation with Lagos as point of sustained port dominance

activity along the Nigerian seaboard was limited; the Gold Coast was then the centre of attraction in West Africa.[2] Benin through its port of Gwato was the leading area of trade, first with the Portuguese[3] and subsequently with the other Europeans who followed them. The Warri kingdom, probably through Ode Itsekiri, also engaged in trade. Bonny was a significant port, and the Europeans had only relatively feeble contacts with the other rivers such as the Brass and Old Calabar. Lagos probably did not exist until towards the end of the period.[4]

1 Babafemi Ogundana, 'Some Elements of Changing Port Concentration in Nigeria', unpublished M.Sc. thesis (University of London, 1966); 'The Fluctuating Significance of Nigerian Ports before 1914', Odu (University of Ife Journal of African Studies) vol. 3 (1967) 44–71.
2 J. W. Blake, European Beginnings in West Africa, 1454–1518 (London, 1937) p. 12.
3 D. P. Pereira, Esmeraldo de Situ Orbis, ed. G. H. T. Kimble (London, 1937) p. 125.
4 A. L. Mabogunje, 'Lagos, a study in Urban Geography', unpublished Ph.D. thesis (University of London, 1961) pp. 26–7.

In the second period, the emergence of the slave trade as a lucrative enterprise attracted many European adventurers to West Africa, and a dispersed pattern of trade developed. In this diffusion, New Calabar and Old Calabar became the leading ports, and the pioneering area of trade in Benin declined. By the middle of the eighteenth century Bonny had taken over the lead in the slave trade, and during the next fifty years it rose conspicuously as the most important port for the trade in western Africa.[1] When the slave trade was made illegal, Bonny switched to the palm-oil trade and maintained its leading position; by the 1850s it had become the most important port for the new trade in all Africa.[2] During the first half of the nineteenth century Lagos developed as the leading port on the Bight of Benin owing to the decline of Whydah (Ouidah) and Porto Novo (in present-day Dahomey), and became the base for recalcitrant slavers. It was ostensibly to stamp out the slave trade that the British annexed Lagos in 1861.[3]

In the imperial scramble for and partition of Africa the British established new interior ports in Nigeria such as Koko, Sapele, Warri, Degema and Opobo (Egwenga), that could serve as centres for the political and economic control of the interior, or as bases for transport developments; Port Harcourt, which operated as a railway base was an example. In 1912 there were fourteen ports in Nigeria, with Lagos handling 41 per cent of the volume of trade, Burutu 15 per cent and Calabar 11 per cent. By 1950 the number of ports was only seven, and Lagos then handled 63 per cent of the total trade. Burutu and Calabar had declined and had been replaced by Port Harcourt (Plate 11), handling 17 per cent of the total trade, whilst Sapele handled 8·5 per cent. Whilst Hirschman's formula gives an index of concentration of 46 for 1912, the figure for 1950 is 66. An analysis of the routing of individual commodities also shows that by 1950 at least 50 per cent of each export commodity was handled in one or other of the two leading ports.[4] In the import sector Lagos had become the outstanding primate port, increasing its share from 42 per cent in 1912 to 77 per cent in 1950.

Since 1950 three new ports have been opened or re-opened: the Bonny

1 *John Adams*, Sketches Taken during Ten Years Voyages to Africa (*London, 1822*).
2 *K. O. Dike*, Trade and Politics in the Niger Delta, 1830–1855 (*Oxford, 1962*).
3 *J. F. Ade Ajayi*, ' *The British Occupation of Lagos, 1851–1861* ', Nigeria Magazine, *vol. 69 (1961); and B. W. Hodder*, ' *The Growth of Trade in Lagos, Nigeria* ', Tijdschrift voor Economische en Sociale Geografie, *vol. 50 (1959) 197–202.*
4 *Babafemi Ogundana*, ' *Changing Port Outlets of Nigeria's Export Commodities* ', Nigerian Geographical Journal, *vol. 9 (1966) 25–43.*

oil terminal, Okrika and Koko. The first of these handles Nigeria's export of crude petroleum; out of the total cargo traffic of 19·3 million tons for all Nigerian ports in 1965–6, 12·3 million tons was crude petroleum handled at the Bonny oil terminal.[1] In the post-1950 period too, with increased direct shipping from the small ports, the ports in the western Delta have gained relatively in the national context, especially in exports. As a result of the mishandling of the Koko port scheme[2] there has been no great diffusion in general cargo flows. Lagos remains the dominant general cargo port, especially in the import trades; Lagos's share of imports fell from 77 per cent in 1950 to 70 per cent in 1960, but recovered to about 74 per cent in 1965.

Factors affecting the structure of the Nigerian port complex

In the search for factors which influence the structure of a port complex, four questions call especially for examination: (1) What factors influence the relative significance of ports in each era of either concentration or diffusion? (2) For each era of concentration, what influences make for the increased significance of the centres of concentration? (3) What factors lead to sustained port dominance? (4) What factors initiate port diffusion? These issues cannot receive detailed treatment here, and only a brief analysis is offered.

Within a port complex, the greater relative significance of any port must rest on certain advantages which operate in favour of that port, and such advantages may derive from one or more of the different factors influencing port significance. These factors can be grouped into four main categories: available port capacity; the location of the port in relation to the potential demand for port services from the hinterland or foreland; the structure and quality of connecting transport by land and sea; and institutional factors in the organisation of trade which influences the flow of goods and the choice of port outlets. These factors and their component elements are in dynamic interplay over time, and are of changing and unequal significance at different ports.

Factors of concentration and consolidation

Factors of concentration and stabilisation of significance are best taken together, because it is the failure of periods of diffusion to counteract the

1 *Nigerian Ports Authority, 11th Annual Report for the Year Ended 31st March 1966* (*Lagos*) p. 93.
2 *Cf. D. Nesterenko and W. R. Tyner, 'Nigeria Plans Port of Koko Expansions',* Dock and Harbour Authority, *vol. 48 (1967) 232–4.*

advantages of a preceding era of concentration that leads to sustained port dominance. An important factor of port concentration is the increasing capital-intensiveness of transport technologies by land, sea and in port facilities. Such developments cannot equally serve all ports, and the selected ports in any one technological era immediately secure a competitive advantage over others. A major factor in sustained and increasing port dominance is the localisation of developments in succeeding technologies at the same port(s). It was the development of railways to serve selected ports (Lagos and Port Harcourt) which rapidly induced a pattern of concentration after the initial imperial diffusion in many parts of Africa.[1] The coming of the railway in Nigeria spelt the doom of the previously dominant Niger–Benue river system,[2] and consequently of the ports served by these rivers. Currently road transport is becoming predominant, and the differential development of trunk-road systems with larger carrying capacities to and from the railway termini has helped to sustain the significance of these ports.

Increasing ship size during this century has been a major influence in the process of port concentration all over the world. The physical facilities for accommodating increasingly large ships can only be provided economically at fewer and fewer points.[3] Such ships, to be profitable, have to call at a limited number of ports. The economic advantages of operating bigger ships have also helped to attract more trade to the ports that can accommodate such ships. The centres of concentration in the continual process of expansion and adjustment to changing demands of transport have thus evolved through more elaborate terminal developments than have the small ports. In Nigeria, there was until the 1950s a deliberate concentration of government investment in port facilities at Lagos and Port Harcourt.[4] Whatever limited development took place in the small ports was almost entirely due to private enterprise.[5] Because of their better terminal facilities and larger traffic flows, Nigeria's leading ports are served by a more extensive network of land transport and shipping services. The natural potential and pattern of areal distribution of development in any country

1 E. J. Taafe, R. L. Morrill and P. R. Gould, '*Transport Expansion in Under-developed Countries: A Comparative Analysis*', Geographical Review, *vol. 53* (1963) 503–29.
2 Gilbert Walker, Traffic and Transport in Nigeria (*London, 1959*).
3 T. Thorburn, Supply and Demand of Water Transport (*Stockholm, 1960*) pp. 142–56.
4 *The basis for this policy was the* Report on the Harbours of Nigeria, Cmd 3205 (*London, 1928*).
5 *As at Burutu; see* '*The Company's River Fleet and Port, Nigeria*', U.A.C. Statistical and Economic Review, vol. 2 (1948) 33–48.

is always far from even, and a factor in sustained port dominance is the regional polarisation of economic growth,[1] the cores being served by the dominant ports and the peripheral areas by the small or the relatively declining ports. One feature of such entrenched contrasts in regional development is the marked traffic potential of the immediate urban environs of the major port.

The changing significance of ports in Nigeria may be seen as a reflection of the duality in the economic potential of the country's three topographical components. Each part, east, west and north, has a core area of high population density and a peripheral area of sparse population and lower development.[2] The importance of Bonny or New Calabar in the Bight of Biafra and Lagos in the Bight of Benin before the twentieth century was dependent on the population potential of their immediate landside area. In the twentieth century most of the country's production and consumption orientated to foreign trade is concentrated in the core areas in the Yoruba west, Ibo east, and central north. Such traffics, either because of the advantage of proximity or of rail transport, found their logical outlet through Lagos and Port Harcourt. These two ports have also grown to be the two leading industrial centres in the country.

Another feature which tends to maintain the dominance of particular ports relates to intrinsic advantages of size. The bigger ports generate increasingly greater pulls than their relatively declining neighbours. Such cumulative growth might be due to external or internal economies.[3] The concentration in big port towns of institutional services for foreign trade, such as banking, commodity markets, forwarding agencies and the like, affords such ports external economies. Such services are not easily developed at new points, and their perpetuation at the larger ports has been a factor of consolidation. In Nigeria, Lagos is not only the political capital but has become the financial hub and the leading business centre.[4] Such a combination of economic functions accounts for the dominance of the port, especially in the import trade. Lagos and Port Harcourt, of all the Nigerian ports in this century, have had the advantages of direct governmental

1 A. O. Hirschman, The Strategy of Economic Development (*New Haven and London, 1958*).
2 K. M. Buchanan and J. C. Pugh, Land and People in Nigeria (*London, 1955*) *pp. 58–99.*
3 *Gunnar Myrdal*, Economic Theory and Underdeveloped Regions (*London, 1963*) *p. 12.*
4 *Babafemi Ogundana, 'Lagos: Nigeria's Premier Port', unpublished dissertation* (*Ibadan, 1960*) *pp. 1–95; and same title*, Nigerian Geographical Journal, *vol. 4* (*1961*) *26–40.*

1   *Nouadhibou (Port Étienne): the iron-ore loading pier*

2   *Dakar: the southern zone of the port*

3, 4 Monrovia: general view of the port (above) *and Buchanan* (below)

5,6 *Abidjan: timber loading in Banco Bay* (above) *and view across the fishing harbour, West Quay and Plateau area of town* (below)

7   *Accra: surf-boat loading prior to 1962*

8   *Takoradi: general view of the port*

9 Tema: the Volta Aluminium Company's berth and alumina store

10 Cotonou: the lagoon entrance, old pier and new port

11   *Port Harcourt: general view of the port*

12 *Lagos–Apapa: the new container berth*

13 *Pointe Noire: general view of port and new extensions*

14 *Mombasa: port and industrial area*

15 *Dar es Salaam: the town and main quays*

administration and, since 1954, of integrated management under the Nigerian Ports Authority.

Factors of diffusion

Major changes in the structure of the economies of the hinterlands and forelands influence the initiation of port diffusion. On the landward side the emergence of a new trade that cannot be served from an existing port leads to the creation of new ports; whilst a relative increase in the traffic potential of the area of influence of an existing minor port leads to its increased significance. The appearance of Bonny and Okrika, for example, in Nigeria's contemporary period of port diffusion has been due to the emergence of the petroleum industry. On the foreland side, upsurges of economic activity in Europe, for instance, during the rise of the slave trade and in the age of imperialism, influenced the dispersal of port functions in West Africa (periods II and IV in Table 10.1). Periods of emergency could have an opposite effect, as during the Second World War when the concentration of shipping was an official policy. Whilst the stabilisation of a technological revolution in transport stimulates a new order of concentration, the transition period from the dominance of one technology to that of the succeeding one is commonly a period of port diffusion and of shifts in port location. In Nigeria, a concomitant factor in the 'imperial diffusion' was the change in transport technologies both by land and sea. On land, the change from dependence on water transport (rivers and creeks) to developed land transport (rail and road) necessitated the creation of new ports with better land accessibility. The change from sail to steam propulsion also had the immediate effect of encouraging the dispersal of port services. Public carrier services were introduced with steamships. This helped a diffusion in the organisation of West African trade by encouraging the small enterprises which could not generate full ship-loads and which could thus not own or charter ships. A frequent effect of an increase in the size of ships serving a complex is port duplication. Outports are established to handle the bigger ships.[1] The outport could, in a succeeding trend towards port concentration, ultimately replace the older port. The opening of the Bonny terminal arose from the possibility of using bigger tankers to evacuate Nigerian crude oil than was possible through Port Harcourt.

Whilst increasing efficiency at developed ports aids concentration, similarly, forms of inefficiency cause diffusion. Port congestion in Lagos

[1] N. J. G. Pounds, 'Ports and Outports in North-west Europe', Geographical Journal, *vol. 109 (1947) 218.*

G

and Port Harcourt in the 1950s, due to saturated capacity, resulted in the diversion of some trade to the small ports. The real threat to sustained dominance is the saturation of physical space for development at a leading port. Such a situation has sometimes led to the establishment of new ports. There is already limited scope for expansion at Port Harcourt. One study has suggested the location of future general cargo facilities for this port around Okrika.[1] Without imaginative land-use planning, the long-term growth of Lagos could also be jeopardised.

Two other factors which influence the tendency to port diffusion can be briefly mentioned. The first is the nature of trade commodities. Bulky or low-value-to-weight commodities, because they cannot stand long overland hauls, tend to encourage a more dispersed pattern of port outlets than high-value goods. This explains the differences in the degree of concentration between Nigeria's export and import traffic. The other factor, which is further touched upon below, pertains to the role of governments, who for political reasons often counter the regional concentration of economic development.

Implications for port development policy

Three sets of problems of practical national significance could arise as a result of the experience of changing port hierarchies: (a) the consequences of relative or absolute port decline owing to an unstabilised port structure, concentration or diffusion; (b) the consequences of port concentration and consolidation; and (c) the problems of evolving a co-ordinated and rational order of ports in the future. The immediate effect of port decline is to lead to redundancy in capital investments, which are largely immovable. But much more significant is the impact of the declining port functions on the economy of the port town and its region. The income from port-orientated employment which previously generated production outside the port system is cut. The need for redeployment of labour commonly leads to emigration to areas of better economic opportunities. The decay of the old exterior ports in the Niger Delta is a case in point. Abandoned warehouses and traces of the floors of demolished trading 'factories' are scenes which remind one of the trade that no longer exists.[2] The previous rich influential centres of local politics and international rivalry have, except for the recent emergence of the petroleum industry, been relegated to economic irrelevance.

1 *Nedeco*, The Waters of the Niger Delta (*The Hague, 1961*) *p. 296*.
2 R. K. *Udo and B. Ogundana, 'Factors in the Fortunes of the Ports in Niger Delta'*, Scottish Geographical Magazine, *vol. 82 (1966) 169–83*.

Along with the tendency of port consolidation there is the regional concentration of other economic phenomena – population, industries, transport facilities, employment opportunities and services of different kinds. Such economic agglomeration leads to potent problems not only at the points and areas of concentration but also on the national level. At the points of concentration there is the abiding problem of phasing port development to match expected growths in traffic. For the port and the port town, there is a need for integrated land planning that takes care of the long-term needs of the port as well as those of industry, services and housing. There is also the regional transportation problem which is increasingly taxed by mounting congestion, and which demands far-sighted planning. Without organised regional planning, haphazard land use could strangle the growth of a port of sustained dominance. In Lagos, the allocation of water frontage to industries, for which such a site is not vital, in the Apapa industrial estate has prejudiced the logical extension of the port along Badagri Creek. It is crucial to the future of the port that very generous allowance be made for long-term port development in the area between Badagri Creek and Lighthouse Beach.

On the national scale, the regional imbalance in levels of development between 'overdeveloped' and 'underdeveloped' areas could have potent social and political effects. Arguments for the initiation of the Koko port development scheme in the 1950s, for example, included a plea for a 'fair share' and 'balanced' distribution of government developments. Such a commonly heard plea for a redistribution of port facilities[1] thus raises the planning problem of guiding the future pattern of port development in order to ensure a rational and satisfactory order of ports.

An attempt to plan port development must answer three related questions: what quantities of port facilities are required, where they are to be located, and at what points in time should they be provided at the different locations. During any one plan period, the options for policy are either further concentration or else a diffusion in the port structure; in other words, either to locate relatively more capacity at the existing leading ports or else at new or subsidiary ports. Whether the development policy adopted is broadly one of concentration or diffusion, a further problem is to determine the degree of change so that the port complex is 'rational' or 'optimal'. But on what set of criteria is one to evaluate a 'rational' port system or a 'satisfactory' degree of concentration or diffusion in a plan period? The adoption of such criteria must rest on the

---

1 *Cf. M. F. Tanner and A. F. Williams, 'Port Development and National Planning Strategy'*, Journal of Transport Economics and Policy, vol. 1 (1967) 315–24.

role ports are thought to play in the economy. The primary role of the port is as a transport terminal, a link between land and maritime transport. The criteria for defining optimality in this case rest on the adequacy of port capacity to service the trade exchanges between the hinterland and foreland. A broader conception of the role of the port in the economy is as a pole of economic development. The major port supports and is a part of a regional agglomeration of economic activities. Port planning from this point of view must take into consideration both the economic activities which the port serves and those it attracts or generates.

A planning principle which emphasises the transport function of the port is the least-cost concept, according to which the objective of transport planning is to minimise the total real cost of satisfying the transport needs of the economy.[1] Applied to ports, the strategy for co-ordinated development is to ensure that the relative sizes of different ports in the complex are such that the complex as a whole is able to function at minimum total costs to the economy. It is important that a comprehensive view of real costs be adopted. Such a comprehensive cost analysis for a port complex should include the economic cost of providing port facilities, as well as the cost of time in ports of ships, cargoes and trucks, and the cost of overland routing of cargoes to and from the different ports.

A recent trend in the economic evaluation of transport projects is the use of cost-benefit analysis.[2] The preferable project is that which maximises the net benefits from an investment. Similarly, the optimal port complex can be said to be that structure which maximises the economic impact of ports on the economy. The size of each port should be such that the sum of the net effects on their region and the economy as a whole is maximised. It is only such an approach that can fully reflect the broader function of the port as a pole of development. The main difficulty and limitation of the usual cost-benefit appraisal of transport projects is the measurement of benefits in money terms.[3] This is also the difficulty in applying the idea of 'maximisation of impact' as a port-planning criterion. The basic question is to quantify what an area gains by having a port. It is also necessary to measure and compare the marginal impact of additional development at different sizes of ports located in areas with different levels of economic development.

As the structure of a port complex is affected particularly by transport

1 H. *Robinson et al.*, The Economic Coordination of Transport Development in Nigeria (*California, 1961*) *p. 3*.
2 A. R. *Prest and R. Turvey*, '*Cost-benefit Analysis: A Survey*', Economic Journal, *vol. 75 (1965) 683–735*.
3 H. A. *Adler*, Sector and Project Planning in Transportation (*Baltimore, 1967*).

technologies, the forecast of likely changes in such technologies is another difficulty in planning rational port development. Such planning is thus especially difficult in a period of major change in transport technology or organisation (as is occurring now on the world scene through unitisation). It is necessary to look beyond the existing port hierarchy to appreciate how the long-term needs of the economy can be satisfied in the new technology, and what influence the ensuing organisation of transport could have on the spatial distribution of economic activities.

Apart from the philosophical and technical problems involved in the formulation of a concerted ports policy, the common deterrent to the achievement of co-ordinated development of port complexes is institutional. The lack of unified authority or of mutual co-operation to harmonise development across related ports, at national or extra-national level, has inhibited rational port development in many areas.[1] In Nigeria, whilst the Nigerian Ports Authority is responsible for development in Lagos and Port Harcourt, facilities in the other ports are diversely owned. There is thus no defined and consistent development policy for the small ports which complement the programmes being pursued for Lagos and Port Harcourt. The lack of co-operation on the extra-national level has resulted in the duplication of port facilities between Nigeria and her neighbours. The advent of unitisation (Lagos already has container facilities [Plate 12]), which for efficiency cannot be controlled rigidly along national lines, calls more urgently for extra-national co-ordination of transport development in West Africa.

Conclusion

Ports function in a widening network of complexes; and related ports influence each other's fortunes. The structure of port complexes develops through a process of *changing port hierarchies*, in which the ranking of ports changes over time and the composite pattern of the complex shows an alternation of concentration and diffusion. Some ports may remain leaders over successive periods of concentration and diffusion. These are the centres of consolidation of port significance. Alternating concentration and diffusion is due to an interplay of many factors, of which the changing technologies of transport by land and sea, and major changes in trade exchanges between hinterland and foreland, are especially significant. The increasing capitalisation of port facilities, and the polarisation of the traffic potential of the hinterlands in favour of particular ports, have been

1 James Bird, 'Seaports and the European Economic Community', Geographical Journal, *vol. 133 (1967) 302–27.*

potent in the consolidation of port significance at a few points of sustained dominance. The important conclusion for port development policy is the need for the integrated planning of port complexes. This conclusion is valid whether the complex is at a local, national or extra-national level. Since the ports in a complex are related, un-coordinated development leads to wastage by the duplication of facilities, and to an expensive transport system.[1]

---

1 See also *Peter J. Rimmer*, '*The Changing Status of New Zealand Seaports, 1853–1960*', Annals of the Association of American Geographers, *vol. 57 (1967) 83–100, and 'The Problem of Comparing and Classifying Seaports*', Professional Geographer, *vol. 18 (1966) 83–91; Dennis E. Kerfoot*, Port of British Columbia: Development and Trading Patterns *(Vancouver, 1966); J. H. Thompson, 'Some Theoretical Considerations for Manufacturing Geography*', Economic Geography, *vol. 42 (1966) 356–65; R. O. Goss, 'Towards an Economic Appraisal of Port Investments*', Journal of Transport Economics and Policy, *vol. 1 (1967) 240–72; McKinsey and Co.*, Containerisation: The Key to Low-cost Transport *(British Transport Docks Board, 1967); National Ports Council*, Port Development *(London, 1966).*

11. Problems of Port Development in Gabon and Congo-Brazzaville

Pierre Vennetier

Dr Pierre Vennetier is Assistant Director of the Centre d'Études de Géographie Tropicale established at the University of Bordeaux in 1968 by the Centre National de la Recherche Scientifique (C.N.R.S.). He previously spent eight years in Congo-Brazzaville working under the auspices of the Office de la Recherche Scientifique et Technique Outre-Mer (ORSTOM). His numerous publications include Géographie du Congo-Brazzaville *(1966) and* Pointe-Noire et la façade maritime du Congo-Brazzaville *(1968).*

The coasts of the Gulf of Guinea and of Equatorial Africa were first opened to European commerce by Portuguese maritime expeditions towards the end of the fifteenth century: the discovery of the Congo estuary dates from 1487. For a long time most of the trade carried on was in slaves bound for the American continent. After the prohibition of the slave trade, it was not until the period of European colonial expansion that the factory stage in trade was superseded; commerce, closely linked to the administrative control and economic development of vast territories, required an increasingly efficient technical infrastructure in terms of river, road and rail links and modern port facilities. It was these new needs which brought about the progressive development of the ports of Libreville, Port Gentil and Pointe Noire (Fig. 11.1). The traffic of these three ocean ports has risen sharply in volume during the past ten years owing to the discovery and exploitation of natural resources, but in each case the structure has remained essentially that of a typical colonial port characterised by a marked imbalance between imports and exports. In Africa links between the coast and the interior have always presented many problems; conditions of geographical situation vary significantly between the three ports discussed in this paper, and inevitably their economic activity is closely linked with the development of new feeder routes. For this reason current railway projects, both in course of study and under construction, are of particular interest in the context of port growth.

Origin and development of port activity

Commercial activity along the coasts of Gabon and the Congo has evolved in an unspectacular manner. For several centuries trade development remained on the bartering level, which called only for brief, irregular contacts between Europeans and the African population. A new stage was reached when settlements were established on the coast, and commercial influences began to penetrate more firmly inland. But when Equatorial Africa was divided between the various colonising powers, the authorities were forced to set up a technical infrastructure capable of handling a growing volume of traffic. This was not accomplished without con-

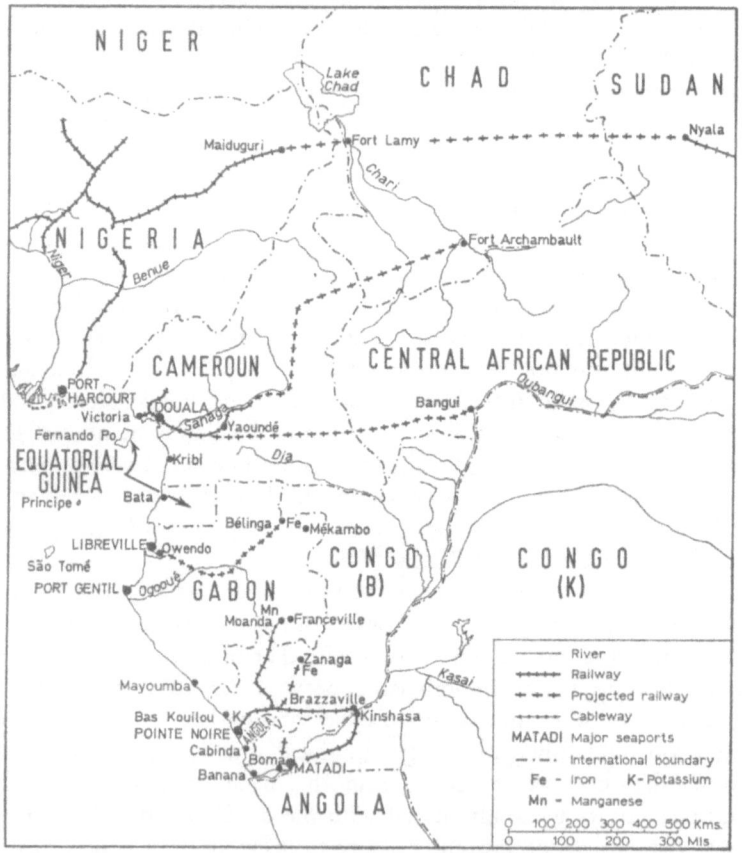

Fig. 11.1   Equatorial Africa: surface transport facilities and major mineral resources

siderable financial and physical problems, and the growth circumstances of trade and transport in Equatorial Africa help to explain the present-day maladjustment of the system to the economic needs of the areas served.

Commercial traffic before the colonial period

From the late fifteenth century to the early nineteenth century commercial traffic did not require any permanent settlements on the African coasts or any firm commercial structure. Slaves and other merchandise, both from the interior of the continent and of local origin, eventually reached the hands of the ethnic groups which occupied the coastal areas, the Mpongoué on the Gabon coast, the Vili on the Congo coast. Transactions took place either on the beach itself, or on board ships anchored a short distance from the shore. The most active trading places were situated to the south of Cap Lopez, the best known being Setté Cama, Mayoumba, Bas Kouilou, Loango, Pointe Noire and Massabi. The native chiefs, whom the Europeans called *mafouques*,[1] monopolised this trade and their share of the profits was invariably the greatest.

This system presented serious problems, however: a ship arriving from Europe was never certain of finding a sufficient cargo, and the buyer was at the mercy of the seller. Traders therefore eventually adopted the factory system as a solution. A factory was a permanent establishment, built near the coast, and run by a factor; it comprised dwelling rooms, a shop and stores, and was erected within a concession bought from the native authorities. Between the departure of one boat and the arrival of another, the factor stocked up with the products that he could gather together or which were offered to him – palm kernels, coconuts, wax, rubber, groundnuts, ivory, etc. – and disposed of the imported goods entrusted to him: alcohol, fabrics, glass beads, gunpowder, arms, etc. In order to increase the volume of trade by controlling a larger territory, a number of factors gradually created a network of secondary stores or branches – called *chimègues* in the Congo – entrusted to responsible Africans whose job it was to stimulate trading activity in the surrounding villages.

Thus, at the moment when colonial penetration effectively began, the coastal regions were dotted with settlements belonging for the most part to companies of diverse nationality in active competition with one another: Britain, Portugal, Holland, France, Spain and the U.S.A. were all involved. Each firm attempted to extend its influence into the hinterland by building factories along the principal rivers (Ogooué, Kouilou, Ouémé), which provided ready-made trade routes, but succeeded in doing

1 *From* mfouka *or* mafouka, *signifying 'chief' in local dialects.*

so only with difficulty, because the indigenous peoples, often hostile to one another, were determined to retain full control of transport within their own territories and to profit from this, for example by enforcing toll dues. Only fragmentary information is available on the tonnage of goods handled during this period, but it is clear that the volume of cargo was very slight; the most active and most advantageously located branches expressed satisfaction when they had dealt with a few dozen tons a year. Price increases between the African coast and Europe often yielded excessive profits.

Many changes were introduced with the establishment of European control in tropical Africa, which involved in this part of the continent the formation of the French Congo and subsequently of French Equatorial Africa. Requirements in materials and provisions for exploratory expeditions for several hundred European administrators, settlers and traders, for the economic and social infrastructure of the country and for the exploitation of mineral and agricultural resources, all entailed a very considerable increase in external trade which could no longer be satisfactorily handled by the methods employed hitherto. Furthermore, the development of urban centres, where most of the Europeans were concentrated, necessarily resulted in a polarisation of commercial trade at certain points, where a minimum level of port equipment became indispensable. These changes were introduced very slowly, but did not assume the same forms in Gabon and in the Congo.

Origin and growth of port traffic in Gabon

The official establishment of a French trading post in the Gabon estuary dates from 1839.[1] The construction of the Aumale fort, which was to consolidate control and help to protect the territory conceded to France, took place in 1843; in 1849 Libreville came into being with the foundation of the village of that name, where forty-six slaves of Congolese origin were settled after having been freed by the frigate *La Pénélope* from a slave ship taking them to America. During the following years the town expanded and developed, but the port facilities remained primitive; vessels anchored in the open roadstead, and the operations of loading and unloading were carried out by means of small boats operating from the French and foreign factories on shore. In 1875 facilities comprised one small landing stage and a fuel depot for the provisioning of ships;[2] two

1 *A treaty was signed on 9 February 1839 between Lieutenant-Commander Bouët-Willaumez and the local ruler, Denis. See G. Lasserre, Libreville, la ville et sa région (Paris, 1958).*
2 *Ibid. p. 78.*

other small stone jetties were under construction. During the following thirty years, when Libreville was the capital of the French Congo (a title ceded to Brazzaville in 1904), there was no consistent demand for a wider range of port facilities, and no major developments were undertaken.

Although the earliest voyages of exploration used Libreville and the Gabon estuary as their point of departure, the natural route of penetration into the interior was obviously the Ogooué river. Port Gentil, which is today the economic capital of Gabon, originated as a small centre of trading activity with warehouses, depots and factories lined up along the eastern side of the Cap Lopez peninsula, which for many years had been a major slave-trading centre (Fig. 11.2). But the growth of Port Gentil, like that of Libreville, has been essentially a twentieth-century phenomenon, due firstly to the 'discovery' of okoumé timber and subsequently to the exploitation of petroleum resources.

Neither the increased number of factories, nor the establishment of a few plantations, nor the efforts of concessionary companies who attempted in vain to launch development schemes, had succeeded by 1920 in stimulating a more vigorous traffic flow through the Gabonese ports. Undoubtedly the reasons for this failure lay partly in the principles upon which these attempts were founded; but environmental conditions were also responsible, particularly in terms of the obstacles which prevented free circulation. The Gabon estuary ends in a cul-de-sac fed only by minor streams, and the Ogooué, interrupted by rapids, does not afford convenient access to the Congo basin. For these reasons, the traffic handled at Libreville was no more than 8,100 tons in 1904 (the first year for which precise statistical information is available), whilst that of Port Gentil – then known as Mandji – was only 9,200 tons.[1] Twenty years later, in 1925, these figures had become 70,000 tons and 166,000 tons respectively.

It was the exploitation of okoumé (*Aucoumea klaineana* [Pierre]) which, especially from 1905, gave rise to an increasing export traffic flow, which in turn led to increasingly heavy and more varied imports. Trade had henceforward to meet new requirements: implements for the timber industry, general equipment for the territory and consumer goods for a population with an increasing amount of money to spend. This new trade structure resulted in an extreme imbalance in port traffic. In 1930, for example, on the eve of the world economic crisis which seriously affected the timber trade, the tonnage of cargo handled at Gabonese ports was distributed as follows: at Libreville, 179,000 tons were exported and

1 *In 1904 imports at Libreville totalled 1,400 tons and exports 6,700 tons. The figures for Port Gentil were respectively 2,300 tons and 6,900 tons.*

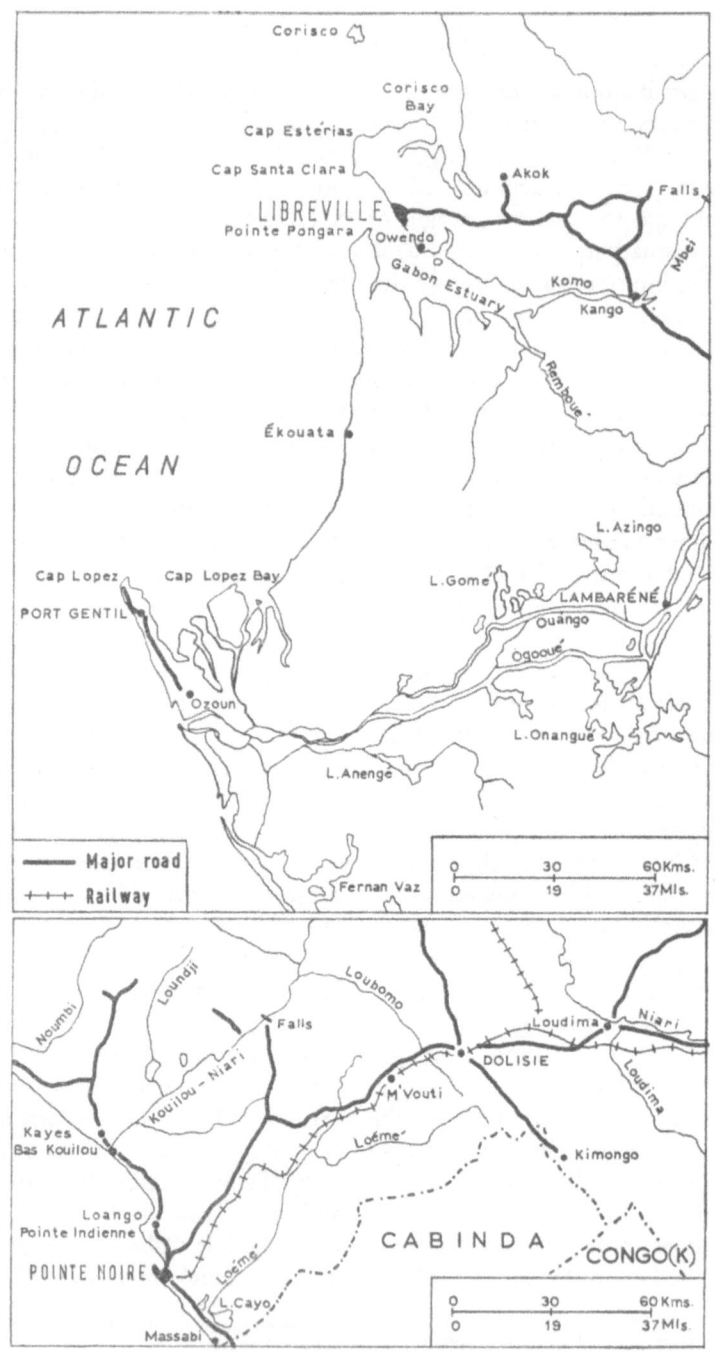

Fig. 11.2   Seaports of Gabon and Congo-Brazzaville

12,600 tons imported; at Port Gentil, 217,400 tons were exported and 16,000 tons imported. The ratio of imports to exports during this period was nearly always 1 to 15 or 20, sometimes even higher. The exploitation of the forest resources enabled Port Gentil easily to draw ahead of Libreville; this was because the areas initially exploited were located in the densely forested region of the lower Ogooué from which river routes converge upon Cap Lopez Bay. In the 1920s the hinterland of Libreville was also the scene of okoumé exploitation, but the port never succeeded in becoming the leading Gabonese timber outlet.

This economic revolution did not have the effects one might have expected upon port development, since the particular methods employed in transporting and loading timber do not require port facilities of an advanced type. Most of the log rafts arrived at a variety of points along the coast, chiefly in the deep indentations such as the Bays of Rio Muni and Mondah, the Gabon estuary, and the lagoons of Fernand Vaz, Iguéla, Setté Cama and Mayoumba. The logs were towed a little way from the shore and loaded directly by means of the ships' own gear, and thus the timber export trade of Libreville and Port Gentil is simply the sum total of the respective traffic of all the open roadsteads which depend on the two ports. This system of operation explains why, for example, port equipment at Libreville in 1925 still consisted of only two nineteenth-century jetties which were, moreover, silted up and almost unusable; the only major improvement in this period was a new wharf, completed in 1930, which comprised an extension of the south jetty, and was equipped with a mobile crane. Apart from timber exports, the volume of cargo traffic handled at Libreville did not exceed 25,000 tons a year, and these very limited port facilities were quite adequate for another quarter of a century. At Port Gentil the situation was very similar; in 1948 the port handled a total of 30,700 tons of cargo imports, Libreville's figure being 15,200 tons.

Origin and growth of port traffic in Congo-Brazzaville

European settlements came later to the Congolese coast than to Gabon. It was only in 1883 that Lieutenant-Commander Cordier obtained two territorial concessions from Ma Loango and one of his vassals, one near Pointe Indienne and the other near Pointe Noire (Fig. 11.2). The explorations of Savorgnan de Brazza and of Stanley had shown that the Ogooué and Congo rivers did not provide convenient arteries of long-distance communication, and it was thought that the Kouilou–Niari route constituted the natural exit from the vast interior region converging upon

Stanley Pool. It was therefore necessary to stake claims on the Atlantic coast. The first administrative post was set up at the head of Loango Bay, partly because the site was healthy (the altitude there is more than 50 m. (163 ft), while at Pointe Noire it is everywhere less than 20 m. (65 ft), but mainly because the royal capital, Boali (or Diosso), was immediately adjacent and was served by an important route from the interior. This route was quickly improved and, as the 'route des caravanes', facilitated head-porterage between Loango and Brazzaville for a period of ten years. Loango was thus the first Congolese port; but, as at Libreville, problems of insufficient capital limited the development of port facilities to two wooden landing stages and a few public and private warehouses. Loango thus remained a lighterage port; traffic rarely exceeded 5,000 tons per annum, and even fell below 3,000 tons after 1897, when most consignments to and from Brazzaville used the newly opened Matadi–Léopoldville railway. Loango continued to stagnate for nearly thirty years, serving only its immediate hinterland, before disappearing completely as a port when the construction of the Congo–Ocean railway and the town of Pointe Noire was begun.

Unlike Libreville and Port Gentil, the port of Pointe Noire was originally conceived as a major element within the modern economic infrastructure of the Congo. Great hopes were founded upon the project – it was to be 'the lung of French Equatorial Africa', the only direct means of access to the Atlantic Ocean from a group of territories covering 2,250,000 sq. km. (869,000 sq. miles). Discussion of possible sites was prolonged for more than thirty years, and fierce controversy was aroused. At one time it seemed that Loango would be chosen as a terminus of the railway, since as an existing and partially equipped township it possessed certain advantages. Advantageous conditions for port development eventually carried the decision in favour of Pointe Noire, which offered a greater depth of water close inshore and less likelihood of rapid silting; swampy conditions rendered the land site much less attractive, however.

From 1921 to 1934 stress was laid on the construction of the railway, which proved difficult and costly in the coastal section. The port operated during this period with the aid of rudimentary facilities, including a temporary wharf, and its chief function was to supply the timber industry with equipment and foodstuffs. Already, however, small quantities of agricultural produce were being exported. Traffic still remained below 50,000 tons, imports regularly exceeding exports. In July 1934 the foundation stone of the modern port, planned by the engineer Blosset, was laid, but the construction was not completed until 1942. By 1939 ships could already berth alongside the wharf and the annual volume of goods handled was more than 100,000 tons. The war put an end to trade expansion until after 1945, when tonnages handled rose gradually to 225,000 in 1948, 280,000 in 1952 and 395,000 in 1955.

An examination of available statistical data reveals a clear contrast between Pointe Noire and the ports of Gabon. Since the beginning of the century the latter were above all timber-exporting ports, and their traffic was largely handled at the open roadsteads. At Pointe Noire, on the contrary, imports exceeded exports until 1953, usually by about 50 per cent. Reasons for this situation are fairly straightforward. On the one hand, immediately after the war, the three colonies served by Pointe Noire profited from a serious effort to establish an economic and social infra-structure which involved massive imports of cement, building materials and machinery, whilst the general rise in the standard of living and the arrival of numerous European settlers entailed an increased demand for imported foodstuffs, textiles and general consumer goods.[1] On the other hand agricultural development made slow progress, and the zone of timber exploitation, less favourably located than in Gabon, did not really expand rapidly until 1950.[2] From the mid-1950s, however, great changes have occurred, which have significantly altered the character of the ports of Gabon and Congo-Brazzaville.

Recent developments and the contemporary situation

In 1966 the total cargo traffic handled in Gabon, both at the ports and at various points along the coast, rose to more than 3,300,000 tons, of which two-thirds passed through Port Gentil and its dependent roadsteads. Traffic through Pointe Noire and its annexes in the same year was 2,350,000 tons.[3] In sixteen years the two states have thus seen their port traffic multiply elevenfold: in 1950 the figures for the Gabonese ports and for Pointe Noire were respectively 200,500 tons and 214,500 tons. This impressive increase has only been achieved as a result of special circumstances and has entailed various important changes.

1 *The minimum daily wage in Brazzaville rose from 3·25 francs C.F.A. in 1939 to 68 francs C.F.A. in 1950. There were less than 4,000 Europeans in Middle Congo, Ubangi-Shari and Chad in 1936, but by 1950 this figure had risen to 16,700.*
2 *On 1 January 1950 timber exploitation permits covered 1,150,000 hectares (4,440 sq. miles) in Gabon but only 175,000 hectares (675 sq. miles) in Middle Congo.*
3 *1967 figures for Pointe Noire show a rise to 2,580,000 tons, of which 2,003,000 tons represent exports.*

Libreville and Port Gentil

For many years the port facilities at Libreville, although rudimentary and outdated, seemed adequate to handle the modest traffic flow, and it was only after 1945 that the need for expansion and modernisation became urgent. In 1954 a new mole was completed, situated to the north of the old wharves, and protected from the swell by a short breakwater parallel to the shore. It provided extensive open stacking grounds and six berths permanently accessible to small vessels drawing less than 2 m. (6·6 ft) of water. Nevertheless Libreville remained a lighterage port where links with the ships anchored offshore were effected by a fleet of barges and lighters. This is still the situation today, since the very shallow waters do not permit the entry of large ocean vessels. Five transit sheds now provide a total of 3,630 sq. m. (4,340 sq. yds) of covered storage space, and this is supplemented by 5,000 sq. m. (5,975 sq. yds) of open stacking grounds; various mechanical aids to cargo handling have also been introduced. A distinction can be drawn between the traffic of the port itself, made up of what is loaded and unloaded at the quayside, and the far more considerable traffic which originates both within the estuary and from the more distant open roadsteads. The former is mainly made up of imports – totalling 117,500 tons in 1966, although there has been a slight downward trend since 1963 – comprising particularly building materials (26 per cent), bulk oils (25 per cent), beverages (16 per cent), foodstuffs and metallurgical products. Total traffic rose to 129,500 tons in 1966 compared with 58,000 tons in 1960.[1] Exports are still relatively insignificant: a few thousand tons of agricultural produce (coffee, cocoa), scrap metal, empty packing-cases, etc. The port is, however, working close to capacity even at this level. Passenger traffic has averaged 4,000–5,000 persons in recent years. The Gabon estuary and the open roadsteads are almost exclusively given over to the export of timber, but as a result of the general reduction in forest exploitation their traffic has sharply declined, especially at Owendo. After having reached a maximum flow of 507,000 tons in 1963, they handled only 326,000 tons in 1966. Thus the maritime trade of Libreville and its associated roadsteads is currently experiencing a decline: having reached 650,000 tons in 1963 total traffic fell to 525,000 tons in 1965 and to only 417,000 in 1966, when some 700 ships entered. If trade with the interior is included, the total figure rises to around 450,000 tons.

The character of the Libreville port region is about to undergo con-

---

1 *Except where otherwise indicated, figures refer to the year 1966. They are abstracted from the* Rapport annuel sur la situation économique, financière et sociale de la République Gabonaise en 1966 (*Libreville: Service National de la Statistique, 1967*).

siderable changes. The need for a deep-water port has become apparent, particularly in view of the probable exploitation of the Makokou–Mékambo iron-ore reserves in eastern Gabon. Soundings revealed the existence of a channel with a minimum depth of 11 m. (36 ft), which would permit large ore-carriers to reach Owendo. A new port is to be built at this point and a considerable sum was allocated for the preparatory work at the end of 1967. The Owendo terminal will offer three deep-water berths, three transit sheds, vast areas for open storage and a timber-stacking zone of 35,000 sq. m. (41,800 sq. yds). An oil berth is also planned, and the foundation stone of a cement works was laid in March 1968 (projected output, 50,000 tons per annum). It is reasonable to suppose that the timber industries will be attracted by the excellent facilities available, and that whilst remaining a mineral and timber port Owendo will gradually see its trade become more varied; but, except during the period of construction and development, traffic is likely to be almost exclusively one-way.

Like Libreville, Port Gentil (Fig. 11.3) is a lighterage port due to shallow water conditions. The facilities available include three moles, of which one belongs to the administration and two to private companies – the Société de Gestion de la Compagnie Française du Gabon (S.G.C.F.G., manufacturers of veneers and plywoods), and the Société des Pétroles d'Afrique Équatoriale (S.P.A.F.E.). Various wharves, stacking grounds (3,900 sq. m. [4,660 sq. yds]), transit sheds (4,500 sq. m. [5,380 sq. yds]), a timber storage area (2,250 sq. m. [2,690 sq. yds]), a mobile crane and a weighbridge complete these facilities, with numerous small boats designed for river navigation. Although the growth of Port Gentil has been encouraged by the timber industry, which still provides a large part of its trade, the discovery of petroleum reserves in the estuary zone and in the offshore zone has recently transformed its activities in a decisive manner. Petroleum exports began in 1957, reached 1,265,000 tons in 1965 and over 3,500,000 tons in 1967. An oil terminal constructed near the tip of Cap Lopez is linked by pipelines to the various fields; storage tanks have a total capacity of 60,000 cu. m. (2,120,000 cu. ft) of oil and loading is effected by oil barges plying to and fro between the shore and the tankers. An oil refinery came into operation in 1968 to serve Cameroun, Gabon and Congo-Brazzaville.

Thus the situation of Port Gentil is not dissimilar to that at Libreville. In the port itself traffic is moderate but increasing: cargo handled in 1961 totalled 130,000 tons, rising to 310,000 tons in 1966. Imports (120,000 tons) in 1966 comprise mainly consumer goods, technical equipment and

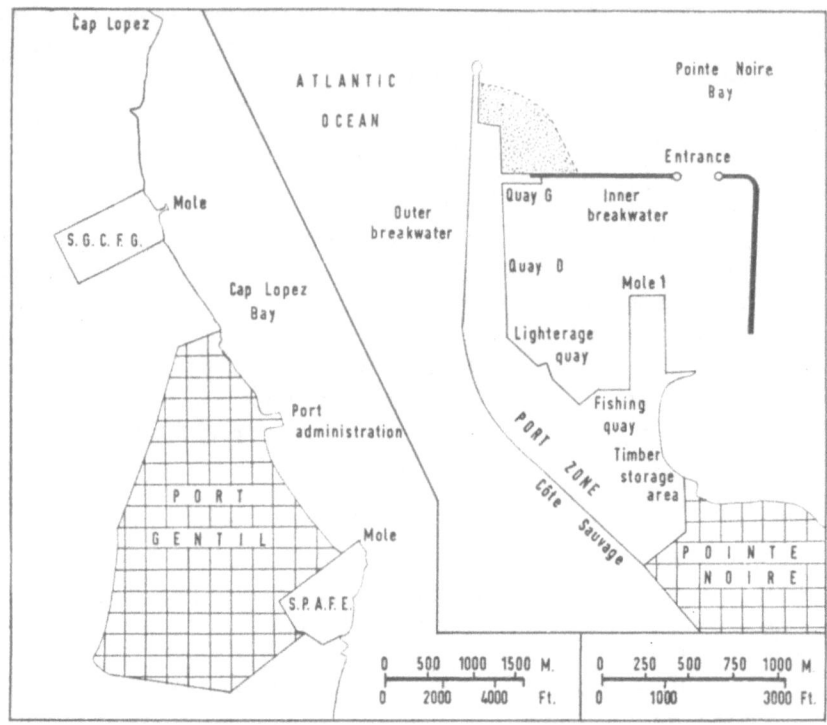

Fig. 11.3   The ports of Port Gentil and Pointe Noire

refined oil products; exports (190,000 tons) consist mainly of timber (particularly the products of the S.G.C.F.G.) and petroleum products. Timber shipped from the open roadsteads added some 236,000 tons to the exports from the Port Gentil port region in 1966, and the Cap Lopez terminal loaded a further 1,360,000 tons of crude oil.[1] Thus Port Gentil now handles a considerable amount of trade and is quite clearly the leading port of Gabon; total traffic has risen from 1,100,000 tons in 1961 to 1,500,000 tons in 1964 and 1,910,000 tons in 1966 (when 850 ships entered). Trade with the interior takes this figure up to nearly 1,950,000 tons. Passenger traffic 2,000 to 2,500 a year, is, however, rather lower than at Libreville.

1 *Crude oil from the Gamba wells, expected to produce 2 million tons in 1968, is loaded directly into tankers by sea-line and does not pass through the Cap Lopez terminal. Exports from Mayoumba, one of the more distant open roadsteads handling chiefly timber, are also recorded separately.*

Pointe Noire

Two principal stages may be recognised in the growth of trade at Pointe Noire (Fig. 11.3; Plate 13): from 1951 to 1962 expansion was fairly steady, and total traffic handled rose from 280,000 tons in 1952 to 885,000 ten years later. This growth resulted chiefly from the prosperity of the timber industry which, in spite of some difficulties, supplied increasing quantities of wood for export, especially limba (*Terminalia superba*) and okoumé. From 1963 onwards a new element became important: the export of manganese ore from Franceville (southern Gabon), mined at Moanda and transported by a cableway and a branch railway to the Congo–Ocean line near Dolisie. Tonnages of ore exported rose from 500,000 in 1963 to 900,000 in 1964, since when the annual total has been about 1,200,000 tons. As a result, the total export traffic of Pointe Noire reached record levels of 1,850,000 tons in 1966 and 2,000,000 tons in 1967 (Fig. 11.4). Imports have also increased considerably, particularly since political independence in 1960, partly as a result of a building boom recalling that of 1946–50; in 1960 total imports stood at 296,000 tons, but rose to 575,000 tons in 1967.

Twenty years ago Pointe Noire seemed doomed to remain a polyvalent port handling mainly small general cargoes, but the port has now taken on a new and more positive character. Today it is primarily a mineral port: manganese ore represents, together with a few minor minerals, some 62 per cent of exports. In the near future this characteristic will be further emphasised with the export of an estimated 800,000 tons of potash from the St Paul workings opened in 1969. Secondly, Pointe Noire is a timber port, handling almost all the timber produced in the Congo (as well as a small part of the production of the Central African Republic). In contrast to Gabon, logs are transported to Pointe Noire by rail and road over a converging transport system which is constantly being extended. Open roadstead loading only takes place at Bas Kouilou, where logs floated 80 km. (50 miles) down the river are shipped (19,000 tons in 1966, 50,000 tons in 1967). The seasonal rhythm of timber exploitation, avoiding the wetter months, is reflected in the port, which has to cope with widely varying monthly totals. Annual timber exports reached 447,000 tons in 1966 and 410,000 tons in 1967. Minerals and timber thus make up 83 per cent of the total maritime export trade.

Pointe Noire is also, though to a lesser extent, a petroleum port. The discovery of an oilfield at Pointe Indienne in 1957 raised great hopes, but the reserves proved to be small, and production, after having reached

120,000 tons per annum, has fallen off to 80,000 tons (1967). A pipeline conveys the oil to the 'Red River Terminal', in Pointe Noire Bay, and a short sea-line is used for loading. To this traffic was formerly added the

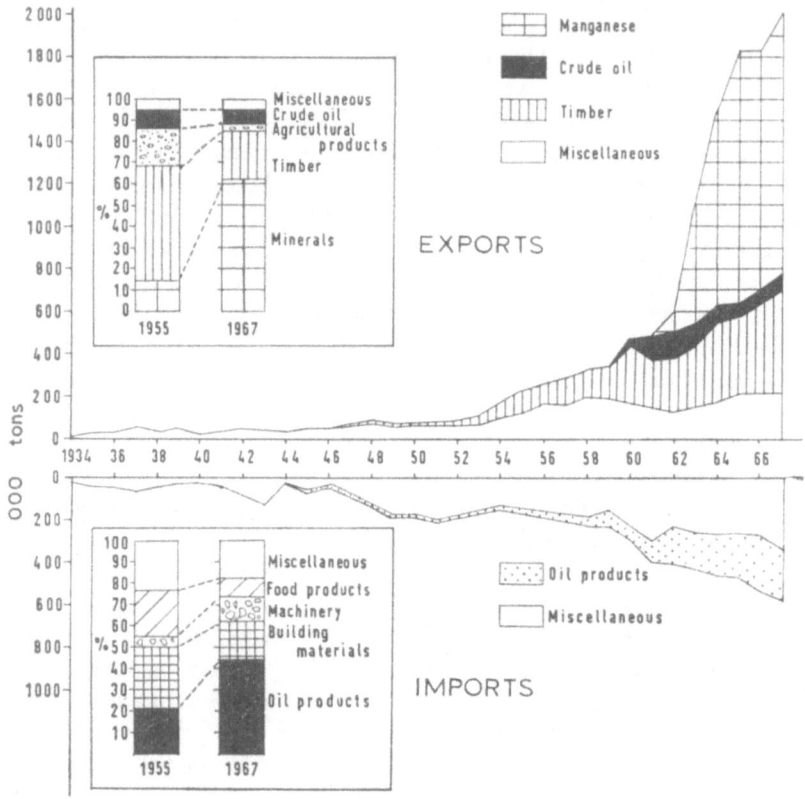

Fig. 11.4   *Growth of port traffic at Pointe Noire*

re-export of refined products bound for Gabon. But imports of hydro-carbons are much more important than exports, and have increased in a spectacular manner (Fig. 11.4) from 100,000 tons in 1963 to 260,000 tons in 1967, and now form almost 50 per cent of all imports. Heavy vehicles and machinery account for much of the increase in fuel oils; and petrol supplies, jet-aircraft fuel and kerosene for domestic lighting are also of increasing importance.

Thus three types of products dominate the trade of Pointe Noire, ores, timber and hydrocarbons, together constituting 81·1 per cent of the total traffic handled in 1966. Other commodities exported include mainly agricultural produce: palm kernels, groundnuts, coffee from the Congo and the Central African Republic, and cotton harvested in the Central African Republic and in Chad. Exports of manufactured goods are very small but include a few hundred tons of palm-oil and oil cake. Sugar exports are increasing rapidly, and reached 61,500 tons in 1967. Congolese production may rise to 150,000 tons by 1970, and of this at least 100,000 tons would be exported through Pointe Noire, destined for a variety of West and perhaps North African markets.

The geographical position of Pointe Noire, at the south-western extremity of a complex system of transport by road, river and rail, which stretches nearly 3,000 km. (1,875 miles) as far as Chad, allows it to play an essential role in the service of three states. An indication of this role is the important position held by cement (65,000 tons per annum) and by various other building materials in the import trade of the port, forming together some 18 per cent of all goods unloaded. Food and drink imports are less important, and consist of merchandise of European origin (salt, condensed milk, tinned foods, etc.) and of traditional foods such as dried fish. The recent establishment of a large flour mill in the Congo and the completion of a brewery at Pointe Noire have drastically reduced imports of flour and beer; cement is also manufactured locally. Pointe Noire also serves as the main entry point for a wide variety of consumer goods, although quantities are small since the market is limited, and a considerable amount of the machinery and other equipment handled is destined for state or private development projects, especially in the fields of communications and the timber and mining industries.

Two ancillary functions of Pointe Noire deserve attention. Since 1950-1 it has been a fishing port, at first relatively insignificant but increasing considerably in importance since 1963. Two small fleets are based in the port, and the annual catch is now nearly 10,000 tons for local consumption. Seasonal tunny fishing has increased recently, the boats calling at Pointe Noire to refuel and to unload; a cold store is equipped to handle 9,000 tons of tunny in transit. Plans exist to exploit marine resources more actively. Pointe Noire is also a passenger port; the number of passengers has always reflected closely the economic and political changes in the hinterland areas. Annual passenger traffic, having increased to more than 10,000 per year, has fallen rapidly since 1959; air competition, reduction in the number of French civil servants and the final departure of many Europeans all help to explain these changes. In 1966 the total number of passengers handled was 5,000.

The colonial heritage of Equatorial Africa, the strength of commercial ties established for over a century, and recent political orientations all influence relations between Pointe Noire and the rest of the world. For a long time France has been her chief trading partner, providing 57 per cent of all goods imported and receiving 62 per cent of all goods exported in 1951; today her role is much less important and the proportion of import and export trade with France stood at 14 and 23 per cent respectively in 1966. In broad terms, Western Europe now takes first place, handling 48 per cent of the volume of imports and 45 per cent of exports; Germany purchases large quantities of timber and Italy supplies oil products. North and South America take second place (40 per cent of exports, 25 per cent of imports by weight), mainly because of manganese bought by the U.S.A. and Canada, and oil products supplied by Latin America. Trade with other African states is small, but imports of oil products from Gabon are increasing. Trade agreements have been reached with Eastern European countries and China.

The remarkable growth of port traffic has been made possible by the installations set up between 1934 and 1942. Favourable factors have included the water site of the port – a considerable stretch of deep, sheltered water protected by a headland – and the overall layout of the port installations, which now include seven berths, extensive stacking grounds and good land communications. The port has been able to handle efficiently the trade in manganese which was not envisaged in the original development schemes, by the adoption of an unusual mechanised system involving the automatic unloading of trucks, open storage and continuous quayside loading. However, problems of congestion have become apparent, particularly in the context of timber exports; logs have been floated in the harbour itself, thereby increasing congestion. The number of waiting days lost by ships became unacceptable by 1964–5, and a new mole was built offering two supplementary berths and several other possible moorings. Potash exports (from 1969) use a wharf developed to the south-east, on the 'Côte Sauvage', the tip of which extends into depths exceeding 15 m. (49 ft). A mechanical loading system using conveyor belts is also in operation here.

Tributary zones and the economic outlook

The functions of Libreville, Port Gentil and Pointe Noire as ports have been the decisive factors in the development of the three urban centres – either directly, by the employment created, or indirectly, by the establishment of industry and commerce with port traffic. This fact is particularly clear at Port Gentil, which has never played an important administrative role.[1] The population of Libreville is today estimated at 65,000 inhabitants, that of Port Gentil at 30,000 and that of Pointe Noire at 100,000; recent increases have been particularly marked at Pointe Noire, which grew from 54,000 in 1959 to 75,000 in 1962. Although precise data are lacking, it is clear that in terms of employment transport is most important (port, railway, lighterage and forwarding companies), followed by the timber and oil industries (the latter particularly at Port Gentil). However, commerce in all its forms employs a greater number. The development of overseas trade during the next decade will thus have important repercussions on the urban economy of the three centres and for this reason it is essential to try to predict the future pattern of port traffic.

Of the three ports studied, Pointe Noire is the only one serving a hinterland which stretches beyond state boundaries. Its tributary area, linked by road, rail and river routes, stretches as far as Fort Lamy in Chad; for a time it even included Katanga, whence some 30,000 tons of copper a year was shipped through Congo-Brazzaville prior to 1960. Since 1963 the hinterland of Pointe Noire has included the Franceville region of Gabon, whence timber as well as manganese is exported by rail via the Dolisie route. The future pattern may, however, be considerably altered by railway projects now being studied.

Since 1962 the governments of Cameroun and Chad have been actively studying the possibility of a Douala–Chad railway link. A preliminary section is under construction from Yaoundé north-eastwards, and surveys of projected extensions to Fort Archambault and to Bangui are under way. Completion of these lines would mean that Pointe Noire would be cut off from a large part of the hinterland from which it draws shipments of cotton and other agricultural products and to which it sends oil products, equipment and foodstuffs. The port would then only serve Congo-Brazzaville and a small part of Gabon, and although its trade would not be greatly diminished it would lose its present diversity, and would become almost exclusively an exporter of minerals and petroleum. This specialisation would increase still further if the Zanaga iron-ore resources were exploited by means of another rail link. Output of iron ore from this source could reach 5 million tons a year.

The prospects for Libreville are not unlike those of Pointe Noire, since the decision to build the proposed Owendo–Bélinga railway has now been taken. By this means annual exports of high-grade iron ore through the

---

1 *Libreville has always been the capital of Gabon. Pointe Noire was the capital of Middle Congo from 1949 to 1959.*

deep-water harbour of Owendo may considerably exceed the conservative estimate of 5 million tons. Timber exports from areas newly opened to exploitation may add a further 1 million tons a year to the export traffic of the Libreville–Owendo complex, which might thus attain, towards 1980, a trade of 8 million tons per annum if industrial activity is attracted to the port area. The hinterland of Libreville would then include almost the whole of Gabon, and would overlap into Congo-Brazzaville. It is not impossible that one day, when the Compagnie Minière de l'Ogooué has paid for its railway, a new, purely Gabonese link might be built for manganese traffic from Moanda to Owendo. For Pointe Noire this would be a very serious loss.

The situation with regard to Port Gentil does not seem likely to change dramatically in the foreseeable future. The timber industry is expanding, although the opening of new zones of exploitation is offset by the closure of older ones in worked-out areas. Oilfields may possibly be discovered in the lower Ogooué region, and an oil refinery will henceforward supply about 500,000 tons of various products, mainly for export to the other states of Equatorial Africa, via Pointe Noire.[1] The essential position of Port Gentil will not be altered: timber and oil will remain the two main elements in its traffic flow.

Port activity in Gabon and the Congo thus presents appreciable contrasts but also many common traits. Geographical conditions and historical circumstances have encouraged traffic at various points along the coast of the Gabon during the past sixty years, while in the Congo trade is almost entirely focused on Pointe Noire. The trade of the ports and the economies of the areas served are in many ways essentially similar: low population densities, low living standards and subsistence economies are still characteristic. Imports of consumer goods remain low, and import traffic is unlikely to increase spectacularly. But the efforts to exploit available resources more fully, to develop interior communications and to establish an essential economic infrastructure are reflected in the importance of imports of equipment, building materials and fuels. In terms of exports it would appear that the 'colonial' function of the ports has been intensified over the past few years: tropical woods, crude oil and mineral ores are being exported at an increasing rate. Profits from these exports are

1 *This position may be altered as a result of the decision in 1968 of Chad and the Central African Republic to withdraw from the Central African Customs and Economic Union (U.D.E.A.C.) and join with Congo (Kinshasa) in the Union of Central African States (U.E.A.C.). Traffic to and from these countries may be diverted from Pointe Noire to Kinshasa and Matadi.*

inevitably far less than if a large part of the products involved were refined or manufactured within the producing countries. Rising port traffic figures merely indicate the movement of raw materials supplying factories and mills located outside Africa. A few, very timid, steps have been made towards industrialisation – veneers and plywood manufacture, the Port Gentil refinery, a sugar refinery and breweries. But shortage of capital development funds and the overall pattern of world commerce do not reveal any hint of profound change in the near future. Libreville, Port Gentil and Pointe Noire seem likely to retain their present character for many years to come.[1]

1 *For further information on the ports of Gabon and Congo-Brazzaville, see J. Bouquerel, 'Le pétrole au Gabon'*, Cahiers d'Outre-Mer, *vol. 78 (1967) 186–99, and 'Port Gentil, centre économique du Gabon', ibid., vol. 79 (1967) 247–74; J. Denis, 'Pointe-Noire', ibid., vol. 36 (1955) 350–68; G. Lasserre, 'Okoumé et chantiers forestiers au Gabon', ibid., vol. 34 (1955) 119–60; M. Roumegous, 'Port-Gentil: quelques aspects sociaux du développement industriel', ibid., vol. 76 (1966) 321–53; P. Vennetier*, Géographie du Congo-Brazzaville *(Paris,1 966); and* Pointe-Noire et la façade maritime du Congo-Brazzaville *(Paris, 1968); D. Hilling, 'The changing economy of Gabon'* Geography, *vol. 48 (1963) 155–65.*

12. The Development and Problems of Port Sudan

M. M. Khogali

*Mr M. M. Khogali graduated in geography at Khartoum University in
1960, and subsequently was awarded a Diploma in Education at Edinburgh
University and an M.A. degree in the University of Wales (Swansea). He
joined the staff of the Department of Geography at Khartoum University
in 1965.*

The Republic of the Sudan is predominantly a land of cultivators and
nomads, importing mainly manufactured goods and exporting agricultural products of which long-staple cotton, cottonseeds, gum arabic,
groundnuts, sesame, hides and skins, and livestock are the most important.
The development of this commercial sector of the economy is post-
nineteenth-century; seventy years ago a subsistence economy prevailed
throughout the Sudan. Surpluses for export were few, and consisted of
such items as gum arabic, ostrich feathers, and ivory. After the establishment
of the Condominium Government in 1898, however, railway lines were
built and a new port developed. These were among the main factors which
led to the gradual increases in the production of both commercial and
subsistence crops. At present the total area annually under long-staple
cotton is about 333,000 hectares (800,000 acres), whilst that under dura
(the main staple diet of the Sudanese) reaches 1 million hectares (2·5
million acres), partly as a result of new mechanised production. In recent
years some light industries have been established, and mining developments
have taken place. Among important industries are cotton textiles in
Khartoum North, and sugar production in Guneid and Khashm el Girba.
These, together with cement and food-processing industries, are meant
to satisfy the home market. On the mining side, iron ore is exported in
small quantities and manganese production is of some importance, although
prospects are uncertain. Port Sudan is the only deep-water port on the
Sudan coast. There are, however, three other very small and specialised
ports: the small jetty at Oseif, about 160 miles north of Port Sudan, which
handles the iron ore of the Fodikwan area; Flamingo Bay, a very small
haven about three miles north of Port Sudan, used by fishing craft and

other small coastal vessels and by sambouks trading across the Red Sea; and the old port of Suakin, still used seasonally by pilgrims leaving for Mecca. Wadi Halfa was formerly the most important of the river ports; but the port ceased to operate in 1964 owing to the construction of the Aswan High Dam, and the town itself has subsequently been submerged. Other Sudanese river ports do not usually handle more than 3 per cent of Sudanese foreign trade (Table 12.1).

Table 12.1

External trade of the Sudan: percentages of the total
volume handled by Port Sudan, Wadi Halfa and the frontier
posts (including Khartoum airport), 1938–67

Year	Port Sudan		Wadi Halfa		Frontier posts	
	Imports	*Exports*	*Imports*	*Exports*	*Imports*	*Exports*
1938	84·7	92·8	13·6	5·6	1·7	1·6
1943	47·8	71·8	40·1	26·1	12·1	2·1
1948	76·2	92·2	18·4	6·6	5·4	1·2
1967	93·7	98·6	0·2	0·8	6·1	0·6

Source: *For 1938–48, the* Reports . . . on the Finance, Administration and Condition of the Sudan (*Cairo, annually); for 1967, the Sudan Department of Statistics.*

Origins and growth of Port Sudan

The foreign trade of the Sudan, from ancient times until the end of the nineteenth century, was carried out by means of caravan links with Egypt and Libya in the north, or by using the ports of the Red Sea coast in the east. The Red Sea, despite the fact that it is characterised by coral growth, has constituted an important water route for many centuries, and various ports have developed at different times along its shores. On the Sudan side these have included such minor havens as Badi and Aythab, whilst Suakin and Port Sudan achieved greater prominence. These various ports have exploited the presence of a series of deep bays and inlets, the cause of the formation of which is not yet fully understood; Crossland attributes the origin of the 'harbours and other fissures in the coral limestone of both coastal and barrier reefs . . . to secondary faulting',[1]

1 C. Crossland, Desert and Water Gardens of the Red Sea (*Cambridge, 1913*) p. 155.

whilst Berry inclines to erosional forces as the major effective cause of their development.[1] In all cases there is no doubt that these harbours, known locally as *marsas*, are relatively deep, though the channels leading to them from the sea tend to be narrow except in certain cases such as Port Sudan where the entrance is comparatively wide.

Suakin, which is thought to have developed originally on its defensive island site at the end of the tenth century A.D., became the only port serving the Sudan after the destruction of the port of Aythab by the Mamelukes of Egypt in A.D. 1428, and it continued to be the main Sudanese port until 1909 when Port Sudan was officially opened. Suakin was a good port for traditional vessels, but it proved impossible to adapt and improve the port to serve the requirements of modern navigation and trade. The break in the reefs which acts as an entrance to the harbour is deep, but it is also narrow for about 48 km. (30 miles) offshore; the water approaches to the port were therefore somewhat hazardous, especially at night. By the turn of the century the town had become congested with old, decaying buildings; and improvements in terms of urban renewal and the provision of safer and more modern navigation conditions and cargo-handling facilities could only have been achieved at high cost. In addition, the water supply obtained from shallow wells was thought to be inadequate for a larger town and port, and the water itself was of poor quality.

Although the disadvantages of Suakin as a port were well known, even before the British occupation of the Sudan, the new administration was given 'the impression that the British naval authorities considered Suakin the most suitable port of entry'.[2] This was why, in the first years of the occupation, it was assumed that port development would be concentrated upon Suakin; and it also explains why the Red Sea railway was built to Suakin and not to Port Sudan. Soon, however, the unsuitability of Suakin became obvious; a serious warning came in January 1904 when the vessel *Ofganistan* was wrecked just outside the harbour.[3] In the same year Captain Kennedy of the Public Works Department advised the building of a new and more suitable port, possibly at Marsa Sheikh Bargout, an inlet used at that time by small local craft, and sometimes by larger vessels for shelter. This inlet, which was subsequently developed and renamed Port Sudan, offered a variety of advantages. In common with all other

1 L. Berry, A. J. Whiteman and J. V. Bell, ' *Some Radiocarbon Dates and their Geomorphological Significance: Emerged Reefs Complex of the Sudan*', Zeitschrift für Geomorphologie *vol. 10 (2) (1966) 135.*

2 Report by His Majesty's Agent and Consul-General on the Finance, Administration and Condition of the Sudan (*Cairo, 1905*) *p. 19.*

3 J. Stone, Sudan Economic Development, 1898–1913 (*Khartoum, 1955*).

ports of the Red Sea the tidal range at this point is negligible; the maximum diurnal variation is under 2·5 m. (7·2 ft), and the greatest seasonal range is less that 1 m. (3·2 ft). Sheikh Bargout had other important advantages, however; the entrance to the harbour is relatively deep and wide, so navigation is less dangerous than elsewhere along the coast and is not restricted to daylight hours. Within the harbour depths ranging from 18 to 24 m. (10 to 14 fathoms) are maintained over a considerable area. The dead coral along the eastern and western shores shelves steeply so that dredging was necessary during the construction of the quay walls. With the possible exception of the squalls sometimes experienced in July and August, the harbour is well protected from prevailing winds and from the open sea by the northern and southern arms of the basin, and by the fringing and barrier reefs. The original inlet also offered plenty of suitable space on the raised coral platform for warehouses and for public and private buildings; and since the pre-port settlement was negligible, demolition and compensation expenses were very slight. Although there was no thorough investigation of the water supply of the new port, it was taken for granted that water supply at Port Sudan would be far superior to that at Suakin. The Governor-General wrote in an annual report that 'an excellent water supply, ample for all shipping requirements and for any conceivable development of the town would probably be available'.[1] For all these reasons Port Sudan was thought to be quite clearly more suitable than Suakin to be developed as the main port of the country.

It was decided that the peninsula between the main part of the harbour and the sea should be used for the building of quays, port offices and warehouses, whilst the town was to be developed on the western side of the harbour (Fig. 12.1). Work in accordance with this scheme was started early in 1906; a branch railway line was built from Saloum station on the Atbara–Suakin line, and the new port was officially opened in 1909. Although growth was initially very slow, Port Sudan established itself in a very short time as the major port of the country and as the main urban and administrative centre of the eastern Sudan; the population rose steadily, and in 1968 reached approximately 110,000 persons. The port of Suakin gradually fell into disuse; as a result of the rise of Port Sudan, Suakin town is now deserted and in ruins. Suakin now has no regular port facilities and it is considered inadvisable for vessels over 100 m. (330 ft) in length or 6 m. (20 ft) in draught to call there. The final blow was given

1 Report   .. on the Finance, Administration and Condition of the Sudan, *1905*, p. *59.*

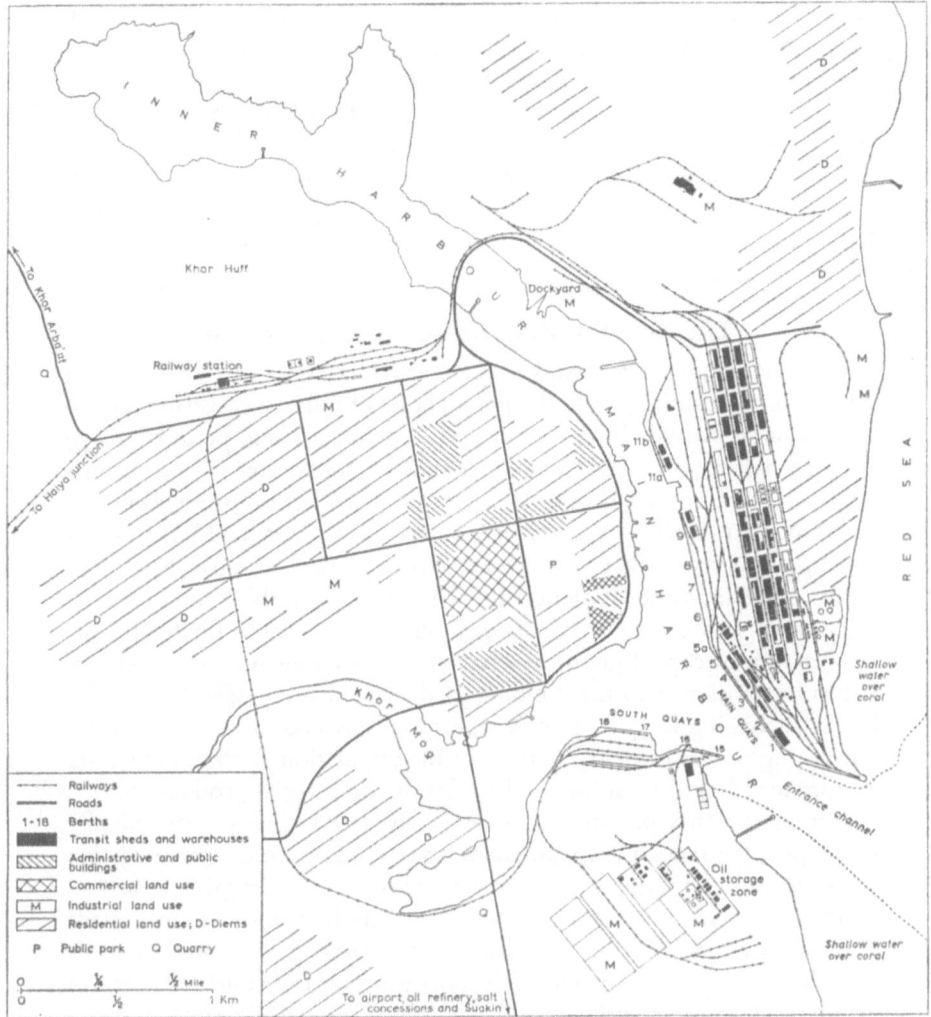

Fig. 12.1   Port Sudan: layout and facilities

to Suakin in 1953 when the railway linking the port with Saloum station was dismantled.

The construction of the railway to the Red Sea, and the development of Port Sudan, were not only essential to the modern economic development of the Sudan but also encouraged the routing of the country's external trade via the Red Sea rather than through Wadi Halfa; today, Port Sudan handles a very high proportion of Sudanese external trade (Table 12.1). This pattern, established during the early years of the century, was disrupted during the Second World War when the proportion of the total external trade of the Sudan handled through Port Sudan fell by almost 50 per cent; in the post-war period trade has once more focused increasingly upon the Red Sea port, so that today less than 1 per cent of all exports and only 6 per cent of all imports are handled elsewhere. In effect, Khartoum airport has become the only alternative Sudanese trade outlet of any significance.

The development of facilities and trade

The development of Port Sudan as a seaport extends over three periods which correlate closely with the main phases of railway construction and of economic growth in the country as a whole. These periods are (a) 1905–24, when port traffic was small and the cultivation of cotton on a large scale had not yet begun; (b) 1925–50, when cotton production was established and traffic increased significantly as a result; and (c) 1951 onwards, when the Sudan witnessed a further period of agricultural and railway development. It should be noted that over these three periods imports as well as exports increased. This is explained, firstly, by the importation of capital goods necessary for development schemes such as railway extensions and the building of dams and bridges, and, secondly, by the increasing imports of consumer goods as a result of the rise in the purchasing power of many sections of the population, such as the Gezira tenants who began to enjoy the benefits of commercial production.

During the first period (1905–24) when port facilities were initially developed, the range of equipment was very limited; only one quay was available, on the eastern side, which, with a total length of 625 m. (2,050 ft), could accommodate five ships simultaneously. It was made clear, however, that further development was envisaged 'as funds become available and the development of the port warrants this course of action'.[1] From the beginning, however, the port was adequately supplied with mechanical equipment in the form of electric cranes and transporters for the handling of coal. Mechanisation was stimulated by the dislike of manual labour on the part of the local inhabitants at that time; the use of temporary labour of nomadic origin is also of interest. In 1909 some 200 men from the Arabian coast were employed in Port Sudan as quay and customs porters, and other workers from the Nile Valley were also employed. Later official reports, however, mentioned that the problem of

1 *Ibid.*

labour supply was gradually solved as more and more nomads came in to seek employment. Some worked only for a short period and then returned to the nomadic life; the majority of workers, however, especially the skilled and the semi-skilled, both from the Nile Valley and from the local area, found that their employment in the port and town was comparatively lucrative and so they settled permanently.

The small size of the port in terms of equipment was reflected in the early period in the restricted traffic flow. The number of vessels entering the port ranged from 195 in 1918 to 699 in 1924 (Table 12.2). On the whole traffic continued to increase, largely as a result of the establishment of peace and security but also through the extension of rail lines and the development of irrigation schemes on a limited scale. The railway to El Obeid was completed in 1911; it passed through the Gezira area where experiments on cotton cultivation were in progress, and it finally linked Kordofan Province, famous for the production of gum arabic, sesame, groundnuts and livestock, with the Red Sea.

The Sudan Government advisers, noting the rapid increase in the traffic of the port, stated in their report for 1924 that 'the existing main wharf has now been worked to its full capacity',[1] and they recommended as a matter of urgency the extension of the main quay northwards for a distance of 70 m. (230 ft) so as to enable five of the larger type of vessel then using the port to be berthed simultaneously. It was also suggested that use should be made of the southern part of the port, previously occupied by the quarantine station, to handle coal and petroleum, thus freeing the northern area to handle general cargo. These recommendations were made in the context of the anticipated expansion of the foreign trade of the Sudan, as schemes for the cultivation of cotton in the Gezira and the Gash delta were about to materialise and the construction of the Haiya–Kassala–Sennar railway was in progress. The idea of physical expansion at Port Sudan was not new, however, for similar proposals had been discussed in the report of the Governor-General in 1905. Adequate funds were not available, either in 1905 or in 1924, and so the advisers did not press for what they regarded as a 'proper course' of ambitious port expansion.[2] The more limited recommendations were, however, put into effect immediately, and by 1928 Port Sudan had a main quay on the eastern side of the harbour with a total length of 695 m. (2,280 ft), designed to handle general cargo, and a southern quay for coal and

1 Coode, Fitzmaurice and Mitchell, 'Proposed Additional Wharfage', unpublished report for the Sudan Government, Oct 1924.
2 Ibid.

H

petroleum. Port officials thought that in ten to fifteen years they might well be looking for another harbour;[1] meanwhile, although traffic in the port increased quite rapidly in general terms, there were important

Table 12.2

Number and tonnage of vessels entering Port Sudan, 1908–24

Year	Number of ships	International tonnage	Net registered tonnage
1908	231	334,251	n.a.
1909	252	383,692	n.a.
1910	251	431,564	n.a.
1911	312	701,110	n.a.
1912	297	700,313	n.a.
1913	314	721,870	n.a.
1914	289	797,278	636,905
1915	202	682,925	n.a.
1916	317	691,899	579,708
1917	288	424,138	349,671
1918	195	315,757	260,728
1919	267	509,498	424,916
1920	308	737,202	610,190
1921	451	1,403,587	1,106,102
1922	452	1,499,935	1,192,660
1923	609	2,265,375	1,765,144
1924	699	2,523,229	1,993,206

Source: *Data collected from* Reports . . . on the Finance, Administration and Condition of the Sudan, *1908 ff.*

fluctuations as a result of the world economic depression in the 1930s and the Second World War (Table 12.3). During the war British naval vessels made considerable use of the port, although civilian merchant vessels were few in number.

1 *Official letter dated 5 Dec 1924, from the Port Manager to the General Manager, Sudan Railways.*

Table 12.3

Number and net registered tonnage of vessels entering Port Sudan, 1925–50

Year	Number of ships	Net registered tonnage	Year	Number of ships	Net registered tonnage
1925	785	2,538,704	1938	1,153	4,004,323
1926	825	2,870,351	1939	922	3,235,331
1927	853	2,986,425	1940	477	1,297,217
1928	922	3,175,692	1941	736	2,369,761
1929	886	3,222,226	1942	582	1,735,551
1930	945	3,491,839	1943	318	871,836
1931	888	3,222,112	1944	418	938,929
1932	808	3,098,307	1945	415	807,034
1933	778	2,920,855	1946	455	1,009,179
1934	886	3,320,622	1947	685	1,946,105
1935	1,181	4,051,804	1948	776	2,171,634
1936	1,148	4,020,134	1949	913	3,054,102
1937	1,174	3,896,216	1950	937	3,049,026

Source: *Annual Reports of the Sudan Railways, 1925 ff.*

After the war the traffic of the port quickly resumed its pre-war level and continued to increase (Table 12.4), partly as a result of long-staple cotton cultivation on the banks of the White and Blue Niles and also of the cultivation of other rain-irrigated commercial crops such as groundnuts and sesame. Gum arabic, which was replaced by cotton as the main export of the Sudan, continued to show a gradual increase although tonnages were small. A further increase in traffic was expected when plans for additional railway lines and for more irrigation works were drawn up. The Sennar–Roseires railway was completed in 1954, and the lines from Rahad to Nyala and from Babanousa to Wau were opened in 1959 and 1963 respectively, the latter mainly for strategic reasons. The Managil extension of the Gezira irrigation scheme was started in 1958, and this has increased the gross cultivated area of the Sudan by some 340,000 hectares (850,000 acres) and the annual area under cotton by a further 100,000 hectares (250,000 acres); more recently a start has been made on the Khashm el Girba irrigation project (Fig. 12.2).

Table 12.4

*Shipping and cargo traffic handled at Port Sudan,
1951–2 to 1967–8*

Seasonal year	Number of ships	Net registered tonnage	Total imports (dwt)	Total exports (dwt)
1951–2	1,025	n.a.	718,247	440,374
1952–3	1,107	3,589,180	671,566	396,484
1953–4	1,202	3,775,869	713,556	528,194
1954–5	1,101	3,325,769	677,746	408,428
1955–6	1,159	3,488,876	789,685	539,027
1956–7	900	2,071,356	839,325	578,620
1957–8	1,206	3,306,817	1,087,716	520,698
1958–9	1,164	3,651,213	858,020	580,471
1959–60	1,312	3,874,672	981,089	706,914
1960–1	1,312	4,056,377	1,077,936	690,400
1961–2	1,285	3,895,247	1,256,065	693,483
1962–3	1,294	4,194,393	1,509,507	933,710
1963–4	1,242	3,920,226	1,791,250	872,350
1964–5	1,166	3,535,446	1,491,906	896,674
1965–6	1,209	3,493,089	1,459,241	899,214
1966–7	1,254	3,661,057	1,469,228	853,782
1967–8	783	2,174,410	1,583,031	919,509

Source: *Annual Reports of the Sudan Railways, 1951 ff.*

A considerable increase in traffic through the port was expected as a result of the completion of these various projects, and the Sudan Government therefore entered into a contract in 1955 to dredge the coral reefs to enable the main quays to be extended by 495 m. (1,626 ft). The actual

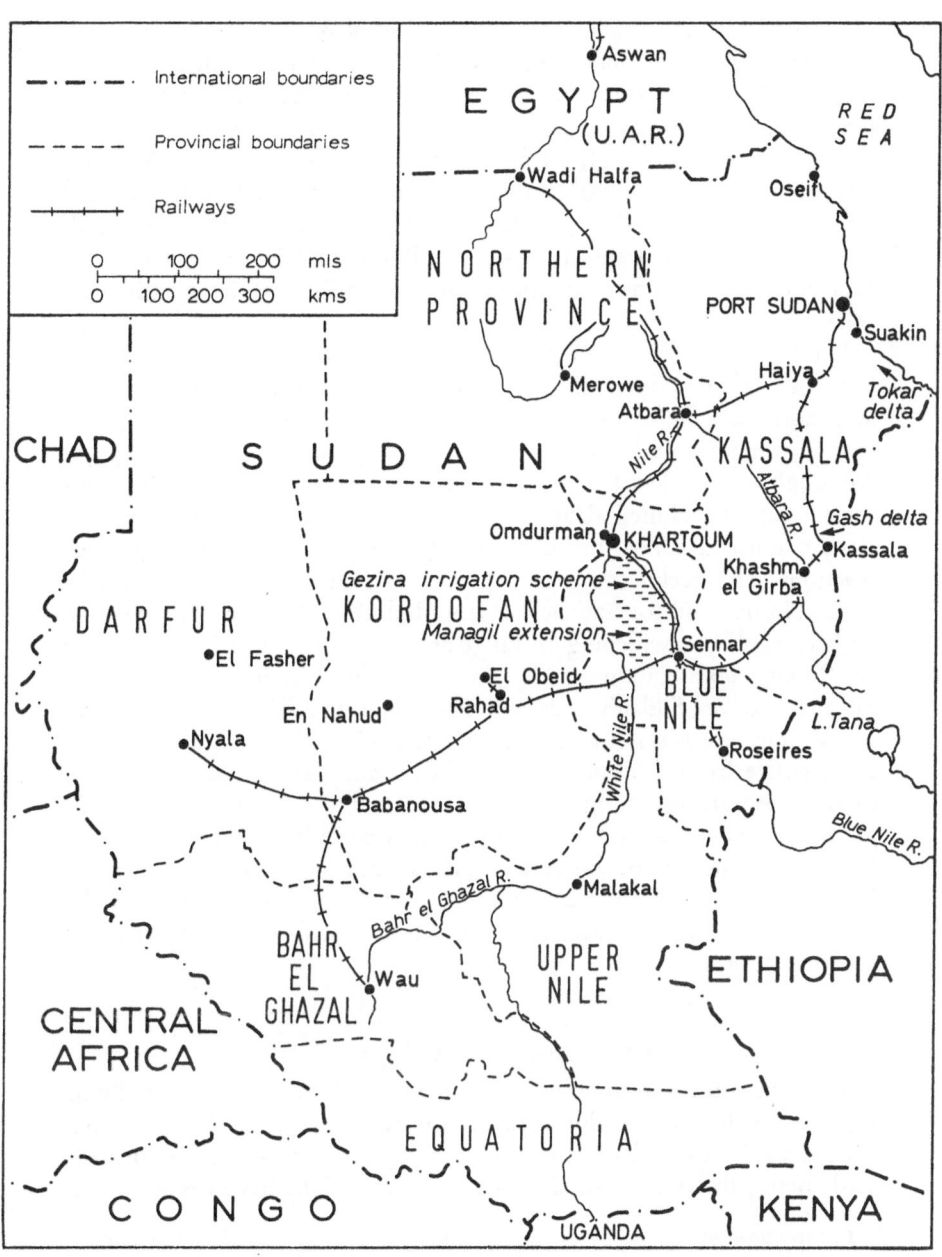

Fig. 12.2   The railway system of the Sudan

construction of the new quay walls was completed by 1960. At present, therefore, Port Sudan possesses two well-developed quays: the main quay, now 1,526 m. (5,080 ft) long, can accommodate ten large ships at one time; the southern quay handles specialised commodities such as petroleum and grains. The total length of the southern quay is 542 m. (1,778 ft) and the depth of water alongside varies between 10·7 and 11·2 m. (35 and 37·5 ft), compared with 8·5 m. (28 ft) at the main quay. Traffic through the port during the period 1951–67 is indicated in Table 12.4, and Table 12.5 shows the volume of the principal commodities handled during the 1962–3 to 1967–8 seasons.

In recent years there has been a significant change in the direction of the Sudan's external trade (Table 12.6). During the period of the Condominium administration the United Kingdom was the country's main trading partner, together with other West European countries, India and Egypt. Although independent Sudan has maintained close commercial links with these areas, the United Kingdom's share of the trade has declined and Eastern Europe, China, the U.S.S.R. and Japan have strengthened their positions. The decline in United Kingdom trade is partly a result of a decline in demand for Sudanese cotton, but also stems from a deliberate Sudanese policy of widening the range of her trade relations. India's rapidly developing textile industry has made her the leading purchaser of Sudanese cotton. Sudan's trade with neighbouring African states and with the Middle East is slight and in relative terms is declining. Sudan imports some coffee from Ethiopia and East Africa and exports dura in return; there is also an increasing livestock trade to the markets of Arabia. In general, however, there has been in recent years a broadening of trading links which can only be to the country's advantage; when the Suez Canal is in operation Port Sudan is ideally located for trade with both the Mediterranean and Atlantic countries as well as with those bordering the Indian Ocean.

Problems of development[1]

One of the most important advantages of Port Sudan is undoubtedly its capacity for further development both as town and port. The town has grown up in an arid region, almost completely devoid of permanent settlements, to become the third largest town of the Sudan with a total

1 *These problems are also discussed in J. Oliver, 'Port Sudan: A Study of its Growth and Functions', Tijdschrift voor Economische en Sociale Geografie (Mar–Apr 1966) 54–61.*

Table 12.5

Port Sudan: volume of principal commodities handled, 1962–3 to 1967–8
(in tons)

	1962–3	1963–4	1964–5	1965–6	1966–7	1967–8
Exports:						
Cotton	185,912	168,572	106,718	132,932	151,601	184,094
Cottonseeds	212,044	101,310	91,269	37,051	52,235	42,351
Gum	43,887	52,124	54,853	55,770	57,069	57;178
Sesame	64,058	85,375	83,367	84,038	80,329	85,208
Dura and maize	69,135	75,356	57,075	99,363	19,762	20,716
Groundnuts	116,368	143,816	162,773	101,737	91,519	122,341
Oil cake	135,767	143,735	149,394	159,596	135,849	179,741
Melon seeds	4,610	7,111	5,877	4,787	5,139	3,363
Cased petroleum	—	458	310	—	1,423	220
Bunker oil	12,785	14,296	16,678	12,736	24,297	24,430
Edible oil	8,768	5,996	13,270	9,591	10,026	10,026
Hides and skins	5,365	5,256	1,886	6,248	7,090	3,984
White oil	—	—	14,646	27,278	15,899	54,430
Dom nuts	18,956	1,738	1,450	989	1,588	1,500
Manganese ore	—	10,533	11,000	11,335	17,264	10,397
Miscellaneous	54,903	59,122	119,925	152,145	181,749	116,949
Transhipment	1,152	737	6,183	3,618	943	2,581
Total exports	933,710	875,535	896,674	899,214	853,782	919,509
Imports:						
(a) Public						
General	712,495	882,802	532,566	500,603	464,096	623,459
Petroleum						
products	544,468	522,750	599,065	640,587	709,941	741,301
(b) Government						
General	145,713	241,165	235,651	106,769	211,576	91,273
Coal	209	6,952	903	314	600	600
Sugar	105,550	126,920	123,314	208,636	81,993	126,327
Transhipment	380	661	407	3,332	23	71
Total imports	1,509,507	1,791,250	1,491,906	1,459,241	1,469,228	1,583 031
Total traffic	2,443,217	2,666,785	2,388,580	2,358,455	2,313,010	2,502,540

Source: *Official records of the Port Office, Port Sudan.*

population of 110,000 persons, including a permanent element and a floating sector. The port is visited annually by over 1,200 ships with a net registered tonnage of approximately 3·5 million. Although the progressive

Table 12.6

Direction of Sudanese external trade by percentage of value, 1954 and 1967

	1954		1967	
	Imports	Exports	Imports	Exports
U.K.	32·5	43·2	20·0	8·1
Western Europe	14·7	20·3	13·5	24·8
Italy	4·9	9·0	4·0	11·8
India, Pakistan, Ceylon, Burma and Malaya	14·0	4·5	10·4	9·5
Egypt	10·4	8·9	4·0	3·9
Americas	1·8	5·8	9·4	6·9
Japan	0·4	0·4	6·6	7·8
Persian Gulf	0·1	—	0·2	0·5
Aden, Saudi Arabia	0·5	1·8	—	2·9
Horn and East Africa	3·2	0·4	1·2	—
Eastern Europe	7·6	0·4	6·2	7·8
U.S.S.R.	0·2	—	1·3	4·3
China	0·1	0·1	7·2	3·6
Others	9·6	5·2	16·0	8·1
	100·0	100·0	100·0	100·0

Source: *Sudan Government, Department of Statistics.*

extensions described above have been considerable (the port now has a total length of deep-water quayage of 2,224 m. (6,858 ft), there is still room for some further extension of the main quays on the eastern side of the harbour; the western side, now lined with public buildings and commercial premises, could also be utilised but would require extensive dredging. Important problems face the port in other respects, however. These include the extensive and remote hinterland, the frequent inability of the port to accommodate all ships calling, the problem of congestion on the quays and in the transit sheds, and the probable inadequacy in the future of the water supply of the town and port.

The problem of its relationships with the hinterland is perhaps the most basic of the various difficulties facing Port Sudan. The Red Sea coast of the Sudan is a desert area, and with the exception of cotton from the Tokar delta and some minerals from the Red Sea Hills it offers very little traffic for the port. This affects Port Sudan in two ways: the first is that the port depends for its supplies of vegetables, fruit and meat on the Nile Valley and the Gash delta. In a country such as the Sudan, where road transport has not yet been extensively developed and rail transport has very little value for perishable products, the prices of fruit and vegetables become comparatively high in the remoter localities and continuity of supply is not assured. The second effect is that the whole of the Sudan, rather than the immediate area, has become the hinterland of the port. But the most productive regions are situated far inland: Khartoum, El Obeid and Nyala are respectively 787, 1,475 and 2,088 km. (492, 922 and 1,305 miles) away by rail from the port. In addition, the productive regions are separated from the port by a vast and partly mountainous desert, which itself offers virtually no traffic. If Atbara, rather than Khartoum, is regarded as the southern limit of the desert, then the desert distance covered by the Port Sudan–Khartoum railway is 480 km. (300 miles). In the case of such long-distance hauls, much of which generates very little traffic, the cost of transport is inevitably high and the potential of the railway as an instrument of economic growth is restricted; trains take at least four days to cover the distance between Nyala and Port Sudan. Conversely, these great distances are clearly among the main factors preventing the use of Port Sudan as an entrepôt for traffic originating in or destined for the Republics of Chad and Central Africa.

Since 1954 the port has been unable to provide immediate berthing facilities for all vessels calling (Table 12.7). In such cases the practice has been either to allow vessels to anchor at Towartit, 12·5 km. (8 miles) to the south, or to resort to double-berthing; the former method causes financial losses to the waiting ships, whilst the latter also involves a variety of other inconveniences. The situation has not been eased by the increase in the length of quays in 1960, nor by the closure of the Suez Canal: before the Middle East crisis of 1967 more ships were calling regularly at

Port Sudan and many spent a longer period of time in the port loading or unloading larger cargoes. After June 1967, however, Port Sudan, having for a long time enjoyed the position of an intermediate port of call, suddenly became a terminus, and the number of ships calling sharply decreased. This was compensated for to some extent by the longer hours

Table 12.7

Some indices of congestion at Port Sudan, 1954–67

Year	A	B	C
1954	1,174	64	10
1955	1,094	63	29
1956	1,074	84	12
1957	1,033	401	226
1958	1,116	111	41
1959	1,308	164	30
1960	1,330	133	16
1961	1,303	164	84
1962	1,298	285	143
1963	1,281	271	164
1964	1,150	206	111
1965	1,200	131	24
1966	1,223	150	46
1967	1,004	65	24

A = Number of ships entering port.
B = Number of ships double-berthed in port.
C = Number of ships waiting at Towartit before entering port.

Source: *Official records of the Port Office, Port Sudan.*

the calling vessels now spend in the port. In normal circumstances, however, when the Suez Canal is open, the problem of the inability of the port to accommodate all vessels immediately on arrival is severe, and in the long term can only be solved by increasing either the available length of quayage or the efficiency of port operation or both. Both these solutions, particularly the first, obviously necessitate the investment of capital which is currently not readily available.

The problem of congestion at Port Sudan is not confined to that of inadequate berthing space. The quays, stacking grounds and transit sheds of the port are usually congested with goods owing to the inability of the single-track railway which serves the port to cope effectively with exports during the peak season or with imports throughout the year. The problem is aggravated by a severe bottleneck between the port and Haiya Junction. Many of the various agricultural products of the Sudan are usually ready for export between November and March; the rain-grown crops, such as dura, groundnuts and sesame, come into the markets as early as mid-October, when the busy season for the railway begins. However, by early January, before these crops are completely evacuated, long-staple cotton is despatched from the ginneries and further strains the transport facilities. Consignments which cannot be transported immediately to the port are kept waiting in the railway stations up country, and to some extent this alleviates potential congestion in the port. Exports also cause problems, in that oilseeds, for example, are stacked on the quayside unprotected, and in the event of rare (but not unknown) summer rain are likely to germinate and be ruined. Imported manufactured goods, on the other hand, are coming into the port throughout the year, and in many cases in larger quantities than the railway is able to forward immediately (Table 12.8). This problem becomes especially acute during construction work on development schemes such as the Managil extension and the Roseires dam when the port handles large amounts of capital goods. Goods sometimes remain in the port warehouses or in open storage for over six months before being transported up country; capital is thus locked up in the goods held in port, and this inevitably increases storage and insurance costs. In addition, items stored in the open air often deteriorate badly in the heat and dust, and from time to time rain adds its havoc. In 1968, £1 million worth of cement was damaged by rain in this way. In addition to raising the prices of commodities, congestion in the port has two other important consequences. The first is that it lowers the handling output of the port. In 1961 the average tonnage handled per hatch-hour was 8·6, rather a low figure compared to the average of 12 tons achieved at both Mombasa and Dakar.[1] However, the Port Sudan average showed a sharp rise during the less congested 1964–5 and 1965–6 seasons, to 17·1 and 17·9 tons

[1] *Sofrerail (Paris), Report on the Sudan Railways, for the International Bank for Reconstruction and Development (July 1963) chap. 9, p. 96.*

respectively.[1] The second consequence of the congestion is that Port Sudan, being unable adequately to serve the economy of its own country,

Table 12.8

Port Sudan: tonnage of goods imported and railed up country during the 1966–7 and 1967–8 seasons

Month	1966–7			1967–8		
	Tonnage imported	Tonnage railed up country	Difference in tons	Tonnage imported	Tonnage railed up country	Difference in tons
July	154,725	117,065	+ 37,660	127,663	90,386	+ 27,277
August	71,259	93,429	− 22,162	115,161	36,874	+ 78,287
September	127,463	100,861	+ 26,602	36,047	87,929	− 51,882
October	36,366	121,292	− 84,926	103,920	109,729	− 5,809
November	94,102	108,597	− 14,494	189,580	114,292	+ 75,288
December	122,220	119,548	+ 2,672	106,719	98,588	+ 8,131
January	160,398	107,587	+ 52,811	138,439	102,792	+ 35,647
February	130,247	112,268	+ 17,979	141,490	90,947	+ 50,543
March	179,808	126,894	+ 52,914	128,621	97,593	+ 31,028
April	110,588	100,619	+ 9,969	172,619	117,342	+ 55,327
May	62,859	95,137	− 32,278			
June	105,687	110,162	− 4,475			

The table shows the total tonnage imported through Port Sudan, the total tonnage transported by the railway to the interior of the country and the difference between the two for each month of the 1966–7 and 1967–8 seasons. The plus sign indicates accumulation of imports in the port, while the minus sign suggests a contribution towards relieving congestion in the port.

Source: *Official records of the Port Office, Port Sudan.*

becomes unattractive as a potential outlet for the foreign trade of the neighbouring countries of Chad and Central Africa.

Although there are no paved roads between Port Sudan and any other part of the country, lorry transport has become quite important in relation

1 *Sudan Railways*, Annual Reports, 1964–65 *and* 1965–66.

to the port in recent years. It is estimated that the number of 6-ton lorries leaving Port Sudan for Khartoum in the six months between November 1967 and April 1968 was 4,560, an average of 760 per month.[1] Lorries of this type usually carry goods such as tea, coffee, spare parts, tyres, etc., that can bear the cost of transport by this means. However, lorry transport is unlikely to offer a solution to the problem of congestion in the port, especially in view of the long stretches of poor roads involved. The problem of congestion at Port Sudan has led many to consider the possibility of reviving the old port of Suakin or even of establishing a completely new port. Such a course of action raises many problems, of which lack of capital is the most important; but at the same time a new port can hardly be justified on economic grounds because 'Port Sudan's difficulties are not the result of its own operation but of the inadequacy of the railway network as a whole to meet the transport demands'.[2] However, the vulnerability of a single port that might be subject to destruction during a war or by the ravages of an epidemic is a cause for considerable anxiety. With a justifiable reluctance to concentrate all their port facilities at one point, the Sudanese Government commissioned the Yugoslav consultants, Pomgrad, to make feasibility studies with a view to the redevelopment of the port of Suakin. In their report of 1964 Pomgrad suggested the development of Marsa Kuwai, about four miles north of Suakin, the construction of a rail link with the Port Sudan–Atbara line, and the location of industry to the north of the new port. Quite clearly the capital needed for such a development was not available and the project has been shelved.

Demands for water supplies at Port Sudan come not only from the port itself and from domestic consumers, but also from industries such as oil refining located in the port. Requirements in each of these sectors have risen as the number of ships calling at the port has increased and as the urban population has expanded. The establishment of new industries in the port would of course further increase the already very heavy demands. The superiority of the water supply at Port Sudan over that at Suakin was one important factor which favoured the choice of Marsa Sheikh Bargout as the location of the new port for the Sudan in 1905. Nevertheless the water supply has been a source of anxiety for the authorities on several occasions. The semi-desert region within which Port Sudan lies experiences a rainfall of approximately 200 mm. (7·9 in.) in the south, but this decreases sharply northwards; the town itself receives an annual average rainfall of

1 *Commissioner, Red Sea Hills, office files.*
2 *Sofrerail, op. cit., p. 99.*

107 mm. (4·2 in.), most of which (74 mm. [2·9 in.] on average) falls in the months of November and December. Wide variations from these average levels are of course experienced. Until 1924 the water for Port Sudan was taken from shallow wells near the small khors of Mog and Huff (Fig. 12.1), but this water was rather brackish and not of high quality. In 1906 preliminary studies were made of the Khor Arba'at and it was agreed that this should become the main source of supply for the new port.[1] Delays occurred, however, and it was not until 1924 that this scheme was put into operation.

Khor Arba'at, with its many branches, takes its water from the Red Sea Hills and drains into the Red Sea some 32 km. (20 miles) north of Port Sudan. The annual flow of the khor takes the form of an irregular series of flushes, each of which lasts only a few hours depending on the amount and duration of rain in the catchment area. The original bed of the khor is thought to be of an old impervious rock, buried under an uncemented layer of debris, gravel and sand. In some parts this permeable layer becomes shallow and allows some water to appear on the surface all the year round. No proper survey of the khor has yet been made and the precise amount of rainfall on the catchment area remains a matter for intelligent guesswork; it is generally agreed, however, that the water of the khor is plentiful and is at present capable of supplying Port Sudan even during the driest months. There is some disagreement as to whether to build a dam or to make deep borings in order to achieve the best results in the future; the idea of a dam has been shelved several times and Port Sudan now depends on boreholes and on a few shallow wells situated some 30 km. (19 miles) to the north-west. Water reaches the town and port by gravity-fed pipes.

Current anxiety concerning the water supply is due to the fact that the water table becomes very low in years of low rainfall, which seem to occur about every six to eight years. The situation presents grave problems when a year of low water table coincides with a year of high water consumption; this happened in 1941 when Port Sudan was a temporary military, naval and air base, and again in 1964–5 owing to the rise in population and the start of building operations on the Shell oil refinery. The Ministry of Works takes the view that a low water table does not necessarily mean that water supplies are becoming exhausted, and officials emphasise the need for more deep boring and great elaboration of the installations. The Shell Company (among others) is not convinced by these arguments, and to avoid future shortages it plans to build a £1 million

1 Report . . . on the Finance, Administration and Condition of the Sudan, *1907.*

condenser. As a result of the increasingly difficult water supply problem other industries may not be attracted to the port and future development may not therefore be as rapid as in the past.

Conclusion

Situated on an arid coast facing (in normal years) one of the busiest sea routes in the world, Port Sudan has the advantage of being an intermediate port visited by a large number of ships, although many of them take on only small cargoes. It is virtually the only port serving the Sudan and it has no competitor for ocean-going vessels. The only effective link between the port and its vast hinterland is an inadequate single-track railway which operates at a low efficiency rate because of the bottleneck between the port and Haiya Junction, the frequent wash-outs along the line during the rainy season, and the shortage of highly skilled labour. Lorry transport is not officially encouraged, but is becoming important during peak traffic seasons in spite of the poor quality of the roads involved. The development of Port Sudan since 1905 has at times been quite rapid and has effectively eclipsed the older port of Suakin, since the new port enjoys a deep and sheltered harbour, adequate water supply and plenty of space for future growth. The modern economic development of the Sudan has resulted from a wide range of factors; but certainly the building of Port Sudan as a modern outlet, and the construction of a direct rail link to the economic heart of the country, were essential prerequisites for the establishment of a viable commercial economy. In this context, however, the Sudan is still far from being near the 'take-off' level. To achieve further progress the railway services need to be considerably improved in order to free the port as far as possible from congestion and thus to increase cargo-handling efficiency. Future expansion of port facilities is planned in due course in order to enable the port to accommodate all ships calling without the necessity of waiting; if this is not done the advantages gained from the intermediate position of the port will be more than outweighed by losses in time and money caused by the berthing delays. It is therefore expected that when the Sudan overcomes the problems associated with her internal transport system and with the port itself, the availability of a more efficient transport infrastructure will at least permit, and possibly encourage, economic advancement on a broader front.[1]

1 *For further information on the Sudan, see Mustafa M. Khogali, ' The Significance of the Railway to the Economic Development of the Republic of the Sudan, with a Special Reference to its Western Provinces', unpublished M.A. dissertation (University of Wales, 1964); R. L. Hill,* Sudan Transport *(London, 1965); Yusuf Fadl Hasan,* The Arabs and the Sudan *(Edinburgh, 1967).*

13. The Emergence of Major Seaports in a Developing Economy: the Case of East Africa

B. S. Hoyle

Lecturer in Geography in the University of Southampton and formerly Lecturer in Geography at Makerere University College (University of East Africa) from 1960 to 1966, and in the University College of Wales from 1966 to 1969, Dr Hoyle has published a study of The Seaports of East Africa *(Nairobi, 1967) and numerous papers on the economic geography of East Africa.*

The progress of a relatively underdeveloped area towards a higher level of economic activity involves a variety of problems and is dependent upon a wide range of factors and influences. Part of the essential infrastructure of modern economic growth is an outline system of surface transport facilities, serving the developing area both internally and externally. In economic terms such a system is an elementary permissive factor, allowing economic interchange to expand and intensify, and at an early stage may also be an active stimulus to development. Within such a basic transport system seaports occupy a strategic place, for they exist to integrate land and sea transport networks. In developing countries this integrating role is especially vital because developing economies tend to be firmly orientated towards overseas markets rather than towards overland trade; because of this often heavy dependence upon overseas trade, the capacity of seaports not only acts as an indicator of the prosperity of the area served but may also directly affect its economic growth by permitting or hindering increased commodity flow.

East Africa is served by five major seaports handling ocean shipping: from north to south these are Mombasa, Tanga, Zanzibar, Dar es Salaam and Mtwara. Although only Mombasa (Plate 14) and Dar es Salaam (Plate 15) are important terminals in a world sense, the smaller ports form a significant part of the overall East African pattern, dealing with commodities that are highly important locally as well as serving large populations. The port group as a whole comprises a hierarchy (illustrated in numerical terms in Tables 13.1 and 13.2) which has developed mainly during the periods of German and British colonial administration in East Africa since the 1880s, and which is a successor to earlier hierarchies which

Table 13.1

Facilities and equipment at East African seaports

	Mombasa	Dar es Salaam	Tanga	Mtwara	Zanzibar
Deep-water berths					
Number	13	3	—	2	—
Total length (ft)	7,690	1,800	—	1,248	—
(m.)	2,344	549	—	381	—
Lighterage wharves					
Number	2	4	2	—	1
Total length (ft)	1,350	1,929	1,250	—	823
(m.)	411	588	381	—	250
Handling points	9	9	9	—	5
Oil terminal	1	1	—	—	—
Dhow wharf	1	1	1	—	1
Transit sheds					
Number	26	16	9	4	2
Area (sq. ft)	1,261,084	873,104	235,867	180,000	62,370
(sq. m.)	113,498	81,100	21,920	16,200	5,794
Stacking grounds					
Area (sq. ft)	880,835	716,934	122,000	140,000	25,000
(sq. m.)	79,275	66,590	10,980	12,600	2,322
Cargo-handling appliances					
Cranage capacity (tons)	638	475	74	85	13
Fork-lift trucks	149	94	17	14	—
Lighters: number	30	65	37	1	24
capacity (tons)	6,945	12,560	7,080	200	4,771
Scammels: horses	33	22	8	9	—
trailers	80	119	27	18	—
Platform trucks	34	45	—	15	—
Pallets	22,634	15,937	1,800	1,413	—

Source: *For mainland ports, East African Railways and Harbours,* Annual
Report, *1968 (Nairobi, 1969) pp. 58–9.*
For Zanzibar, miscellaneous local sources.

developed along this coast in the past in different circumstances. Heavy capital investment in port and railway facilities in the twentieth century, as a result of modern technological requirements, thus appears to have crystallised a formerly fluid port pattern; but, as will be shown, the balance within the existing hierarchy is constantly changing in response

Table 13.2

East African seaports: shipping, passengers and cargo dealt with in 1968

	Mombasa	Dar es Salaam	Tanga	Mtwara	Total
Shipping					
Ships entering port	2,114	1,003	520	195	3,862
Net registered tonnage	7,988,329	3,635,060	1,906,365	632,767	14,162,521
Passengers					
Disembarked	16,459	17,537	1,326	862	36,184
Embarked	22,238	15,603	472	1,254	39,567
Total	38,697	33,140	1,780	2,116	75,751
Cargo					
Imports:					
Dry cargo	930,902	434,670	46,537	30,229	1,442,338
Petrol and oils in bulk	2,346,364	899,839	19,136	17,413	3,282,752
Total	3,277,266	1,334,509	65,673	47,642	4,725,090
Exports:					
Dry cargo	1,438,922	585,058	180,022	125,833	2,329,835
Petrol and oils in bulk	791,649	208,167	—	—	999,816
Total	5,507,837	2,127,734	245,695	173,475	8,054,741
Total (incl. transhipment)	5,528,341	2,140,618	246,633	173,893	8,089,485

Source: East African Railways and Harbours Administration, Annual Report, 1968. Cargo figures are in metric deadweight tons. The port of Zanzibar is not included in the table owing to unreliability of data.

to a wide range of influences. In this chapter an attempt is made to elucidate the conditions which permitted and encouraged the emergence of the modern East African port group. The main factors that appear to have been involved are discussed; taken together, these factors form a broad conceptual framework which, although arising from the study of a specific port group, may be relevant to other groups elsewhere. Factors which help to explain differential growth within the group are emphasised, as well as factors affecting the emergence of the group as a whole.

Environmental factors

Numerous problems have been associated traditionally with the physical nature of the African coasts and have often been adduced as a major reason for the delay in the growth of modern seaports and in the wider economic development of the tropical regions of the continent. Problems arising from the infrequency of good natural harbours, the absence of large navigable rivers as arteries of penetration, from swampy deltas, coral reefs and surf were certainly important locally in many areas, but have not hindered the provision of modern port facilities required within a wider context. Certainly it is no longer possible to contend that lack of proper port facilities is delaying African economic expansion in broad terms, as most stretches of coast are now equipped with good harbours, natural or artificial, and with a reasonably wide range of modern port equipment. Factors associated with the physical environment have strongly influenced the emergence of the modern seaports of East Africa. The relative strength of their influence has varied through time so that it is possible to point to major physical influences on the maritime side in the pre-twentieth-century eras of development, whilst the more recent growth of the seaports has been closely associated with physical conditions in the hinterland. In addition, the physical nature of the immediate coastal area has been influential in different ways at different times.

On the maritime side the basic environmental factor in the development of trading activity on the East African coast has been the influence of the monsoonal wind system – the seasonal reversal of winds (Fig. 13.1) which brought traders and invaders from southern Asia to the shores of East Africa from early medieval times onwards, and which thereby included East Africa within the trading circulations of the Indian Ocean as a whole. In physical terms this factor is now of very minor importance, since the dhow trade which traditionally depended upon it has declined almost to insignificance;[1] but it is relevant as the basic maritime physical factor affecting the East African port pattern in the nineteenth century, the immediate precursor of the pattern of today. On the landward side the influence of the physical environment of East Africa upon port growth is seen mainly through physical controls on production and movement. Fig. 13.1 shows the overall geographical disposition of areas in East

---

1 *For a description of the dhow trade in relation to East Africa, see D. N. McMaster, ' The Ocean-going Dhow Trade to East Africa'*, East African Geographical Review, *vol. 4 (1966) 13–24. Dr McMaster also provides a useful summary of the meteorological conditions affecting early navigation in the north-western Indian Ocean.*

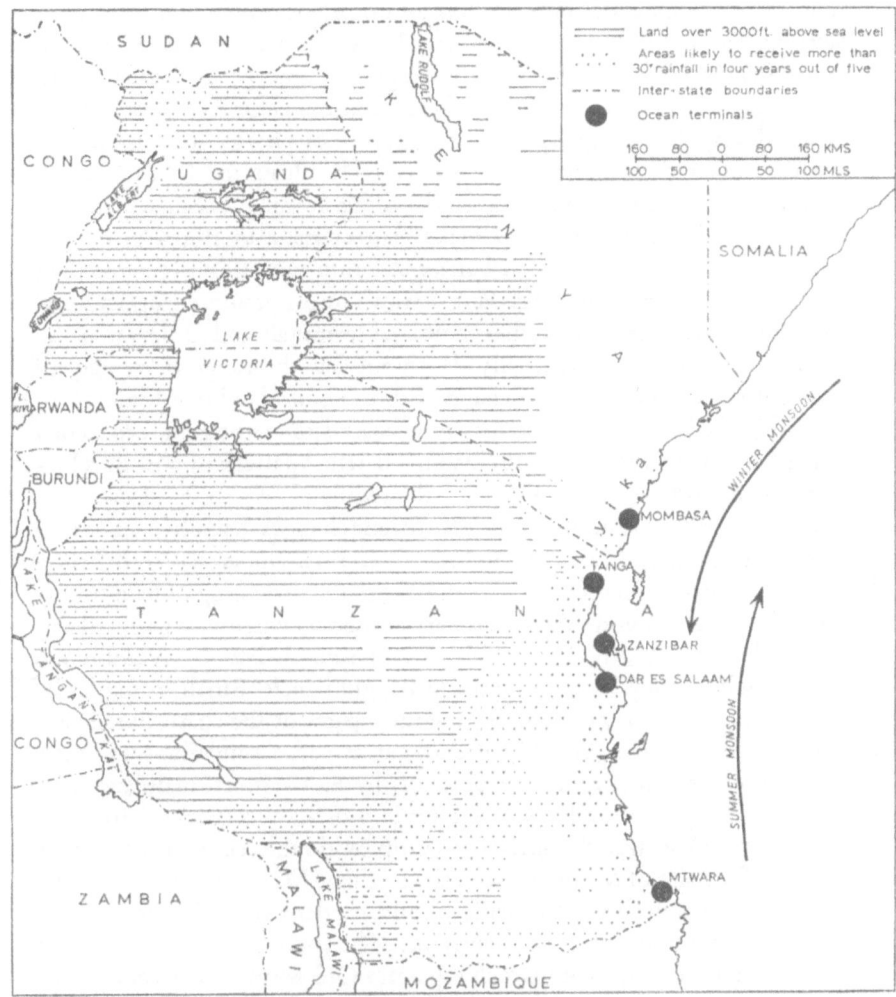

Fig. 13.1 Some environmental factors influencing the emergence of East African seaports. Relief and rainfall data are based on maps accompanying E. W. Russell (ed.), The natural resources of East Africa (Nairobi, 1962).

Africa receiving over 76 cm. (30 in.) of rainfall in four years out of five, and of areas over 900 m. (3,000 ft) in altitude. Within the relatively high, relatively well-watered regions favoured with both these attributes are found the main zones of economic production for export, of population concentration and of import consumption; through the economic development of these zones the physical environment of East Africa has exerted a marked degree of control both upon the emergence of the modern seaport group as a whole and upon the differential pattern of development discernible within the group. The importance of physical conditions in a broader rather than a local sense is effectively illustrated by the port of Mombasa; the immediate hinterland of this leading ocean terminal is part of the dry thorn-scrub or *nyika* characteristic of much of eastern Kenya and forming an effective barrier to communication in the past, but the wider hinterland includes East Africa's most favoured agricultural areas. In contrast, the three other mainland terminals and Zanzibar are all more favourably situated locally, but none has the range or variety of environmental conditions associated with the hinterland of Mombasa.

Within the immediate environs of the coastal zone, the geomorphology of the East African coast has moulded the detailed pattern of port development. The most significant feature is the existence of a series of drowned river valleys or rias resulting from Pleistocene changes in relative sea-level.[1] These rias vary considerably in size; their rock floors are well below present sea-level, and they clearly result from the drowning of land valleys carved by river erosion. Many are characterised by narrow, winding deep-water channels, partly blocked by submarine shelves of recent coral growth where they pass through the fringing reefs. Small examples of rias of this nature may be found along most parts of the coast; in the past, many of them have probably been used at some time for small-scale trade purposes. The modern seaports of the East African mainland have utilised the larger examples, which offer in most cases a spacious stretch of sheltered deep water. A somewhat complex double ria forms the basic feature of the site of Mombasa, and the fact that Mombasa has been a seaport of significance in both medieval and modern times is largely attributable to its possession of two harbours. The town and most of the port facilities stand on an island which separates Old Mombasa harbour – the medieval port – from Kilindini harbour where modern deep-water facilities have been developed. The harbour of Dar es Salaam is comparable to Kilindini, although somewhat smaller and more difficult of entry; and Mtwara harbour is similarly part of an extensive and deep ria system. At Tanga, the economic utilisation of the ria available there has been limited by shallow water conditions. Zanzibar harbour, which is also rather shallow, is not a ria but an area of water sheltered by small islands and

1 H. L. Sikes, ' *The Drowned Valleys on the Coast of Kenya*', Journal of the East African and Uganda Natural History Society, *vol. 38 (1930) 1–9.*

coral reefs, whilst the promontory on which the town is sited itself affords some protection from the strong monsoonal winds. Physical site factors thus favour particularly the ports of Mombasa, Mtwara and Dar es Salaam, but operate against the continued growth of Tanga and Zanzibar. A ranking of the seaports in terms of site conditions and site potential does not accord closely with a ranking in terms of facilities and traffic flow; this is a measure of the importance of conditions of situation – physical, economic and political – operating within a far wider framework.[1]

Historical precedents

In historical perspective the emergence of the modern East African seaport group is seen as the most recent stage in an extended period of trading activity stretching far back into the obscurity of the earlier centuries of the first millennium A.D. For the greater part of this period the East African coastlands were for trading purposes little more than part of the western shore of the Indian Ocean.[2] Within the not inhospitable environment of the coastlands, successive generations of traders and invaders brought by the monsoons built up contrasting hierarchies of seaports; and the growth of port activity in East Africa thus illustrates the concept of changing port hierarchies and also emphasises the need to view land and sea influences upon port development in the context of man's technological achievements in a given period.[3]

Documentary or archaeological evidence of early trading patterns on the East African coast is generally slight, but it is possible to perceive the character and extent of East African trade at certain periods and the distribution and activities of the ports through which that trade passed. The earliest surviving description of the external trade relations of East Africa is to be found in the *Periplus of the Erythraean Sea*,[4] a guide to the commerce of the Red Sea and the Indian Ocean written in the first century A.D. Ports known to the author of the *Periplus* included Mombasa and probably the island of Zanzibar, and the impression is given that East

---

1 *Further details of environmental conditions and other aspects of the port group are available in* B. S. Hoyle, The Seaports of East Africa (*Nairobi, 1967*).
2 *The phrase is taken from Sir Charles Lucas,* The Partition and Colonisation of Africa (*Oxford, 1922*) *p. 9.*
3 *This section is based upon* B. S. Hoyle, 'Early Port Development in East Africa: An Illustration of the Concept of Changing Port Hierarchies', *Tijdschrift voor Economische en Sociale Geografie, vol. 58 (1967) 94–102.*
4 W. H. Schoff (*ed.*), The Periplus of the Erythraean Sea (*London, 1912*). *A new edition of this work is to be published by the Hakluyt Society.*

African trade was at that time a rather tentative offshoot of the wider and firmer circulations of the north-western part of the Indian Ocean. The activities of the medieval seaports of East Africa represent an intensification of this pattern of commercial exploitation, stimulated by new external influences arising from the much fuller development of Islamic culture and trade, particularly in the fourteenth century. The emergence of Kilwa (Fig. 13.2)[1] as a major medieval seaport, controlling trade along the East African coast from Malindi to Mozambique, was due mainly to three factors: a defensive island site, the convergence upon the port of major inland trade routes, and the control exerted over supplies of gold reaching the tributary port of Sofala from what are now Rhodesia and South Africa. In Arab terms, East African commercial prosperity in the medieval period was linked with the great religious revival which brought East Africa for the first time firmly within the Islamic world. The prosperous Indian Ocean trade system thus established by the Arabs largely disintegrated under the impact of the Portuguese, whose invasion of the area at the end of the fifteenth century was chiefly motivated by the desire to control maritime trade. On the landward side, a contributory disruptive factor was the continued southward penetration of the Galla along the coast in the seventeenth century, pushing back the northern frontier of the Bantu-speakers almost to Mombasa; this hastened the abandonment of many port sites and concentrated trading activity on defensive islands such as Mombasa, Lamu and Pate. After the final withdrawal of the Portuguese early in the eighteenth century, the seaports of East Africa lay 'in a backwater, aloof from the world and ignored by it'.[2]

The complex historical processes of the nineteenth century produced in East Africa a major change in the geographical relationships of the coastal ports with the interior hinterland and with the rest of the world. A strong renewal of Arab trading activity and the reappearance of Europeans were the basic factors underlying the period of exploration, proselytism and trade which led to the establishment of European colonial rule. During the middle and later years of the century the port and town of Zanzibar emerged as the principal entrepôt and outlet for East African trade (Fig. 13.2).[3] With a valuable strategic and commercial situation and the best fresh-water supply on the East African coast, Zanzibar was able to maintain this predominant position until it was effectively undermined by

1 *The author is indebted to Miss Alison Smith and Mr James Kirkman for advice in
   connection with sections (a) and (b) of this diagram.*
2 R. Coupland, East Africa and its Invaders (*London, 1938*) *p. 71.*
3 *A detailed contemporary account of Zanzibar at this period is Sir R. F. Burton,*
   Zanzibar: City, Island and Coast (*London, 1872*).

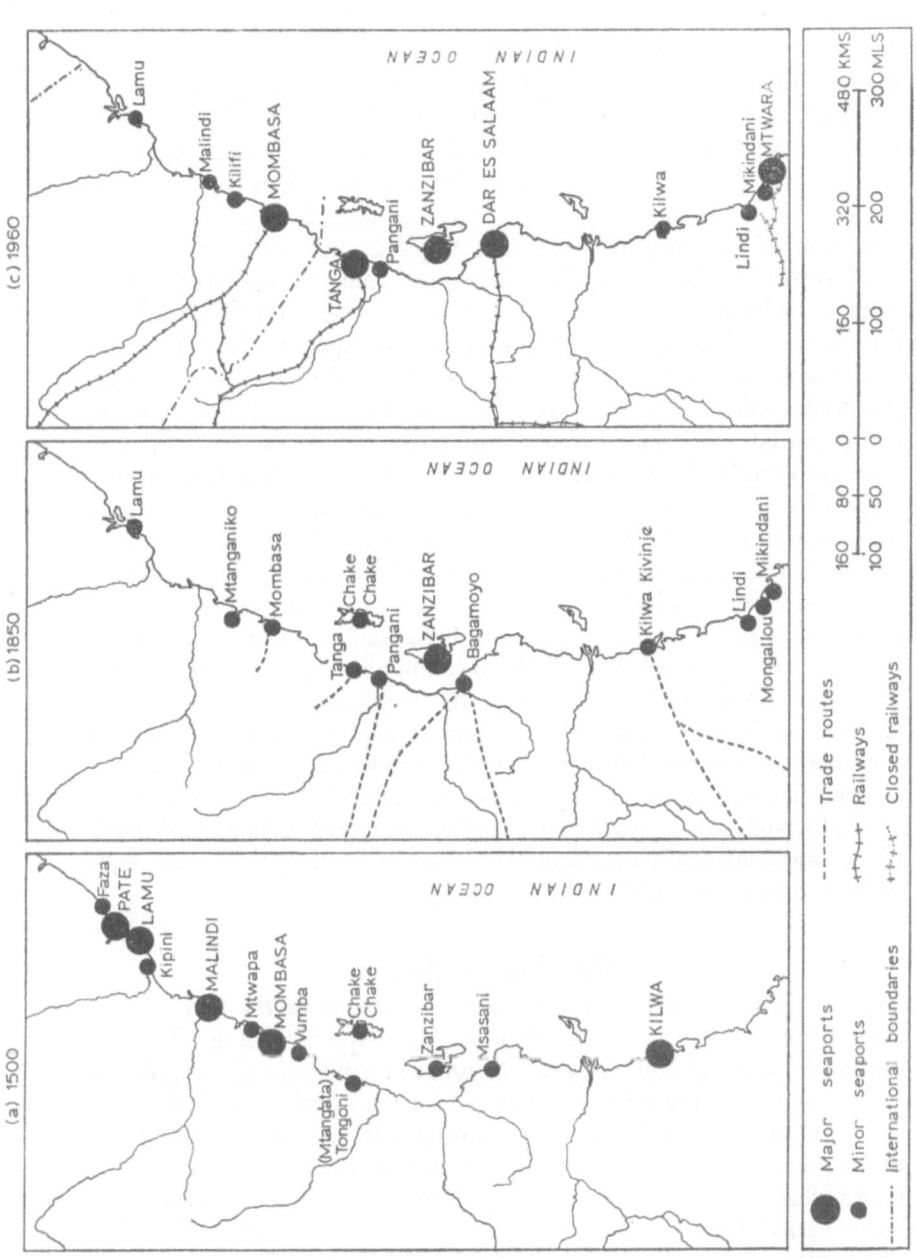

Fig. 13.2 Some examples of East African seaport hierarchies

the establishment of railway links between the mainland seaports and their interior hinterlands. The advantages which the deep, sheltered harbours of Mombasa and Dar es Salaam now hold over the shallower, smaller and more exposed port of Zanzibar only became important towards the end of the century as technological, political and economic influences changed; earlier, in mid-century, the commercial opportunities of Zanzibar were much better, for the port lay opposite a broad section of the fertile coastal belt from which led the easiest routes of access to the interior. By utilising these natural advantages in a period of rapid change, and by developing small mainland outposts at Bagamoyo, Saadani and Pangani, Seyyid Said (the trader-ruler of Zanzibar from 1832 to 1856) was able to bring the entire East African coast from Mogadishu to the Rovuma river, together with the hinterland as far as it was then known and developed, within the range of a single political and economic system – albeit a tenuous one. The invasion of Arab and Western commerce into East Africa, from its base at Zanzibar, was soon followed by other invasions of missionaries and explorers, by the gradual abolition of the slave trade and by the political demarcation of the mainland between Britain and Germany in 1886. As the pace of change quickened, the Arab commercial system foundered and the focus of port activity moved to the mainland, to the deep-water harbours with modern facilities including feeder railways. Attempts to perpetuate the role of Zanzibar as a major seaport have largely failed, but an appreciation of its role as a precursor of the modern port group and as a filter through which passed so many of the change-stimulating influences of the nineteenth century is essential to an understanding of the emergence of the modern terminals.

Technological changes

The emergence of successive hierarchies of seaports on the East African coast demonstrates that port development is an expression of the changing interrelationships of land and sea. The precise form which this expression takes at any given point in time is itself a function of the stage reached in man's technological advancement. Throughout the greater part of the development of port activity on the East African coast the rate of technological change has been comparatively slow and the importance of new technological factors correspondingly slight. The types of Arab sailing dhow traditionally used for the journeys to East Africa[1] were finely adjusted to the monsoonal wind regime which they exploited and to the

1 McMaster, *op. cit.*

modest harbours which they utilised, but they represented nevertheless a relatively elementary technological stage.

The redevelopment of Arab trading activity and the re-entry of Europeans on the East African scene in the nineteenth century coincided with a major period of technological change on a world basis arising from Europe's industrial revolution. A basic factor in the new geographical situation of East African seaports in the later nineteenth century was the Suez Canal, opened in 1869, which channelled major sea-lanes through the north-western Indian Ocean, rescued the East African coast from commercial oblivion and hastened the replacement of the Arab thalassocracy in the Indian Ocean by a far wider trading system based upon European technology. Two other navigational factors were of paramount importance: the change from sail to steam propulsion and the rapidly increasing size of ships. The combined effect of these two factors was extremely important, for together they made the utilisation of deeper and more spacious harbours both possible and necessary. Some misgivings were expressed in the 1880s as to the feasibility of navigating vessels larger than the dhows into and out of the somewhat restricted entrances of the major East African deep-water harbours; but with careful lighting and buoying this problem was overcome, the manœuvrability of the new ships was effectively demonstrated,[1] and the small harbours used by the dhows fell increasingly into disuse.

These technological changes came into effect in an East Africa in which European economic and political penetration was proceeding rapidly ahead but where Arab trading systems and political interests were sharply contracting; this was partly because of the Arabs' inability to withstand European competition (especially after the Europeans had effectively abolished the slave trade, the traditional linchpin of East African coastal commerce), but also because of Arab reliance upon an outmoded technology. The result of these developments was a relocation of the major seaports of the East African mainland and the emergence of a new port hierarchy based on the deep-water harbours hitherto almost completely unutilised. Around the turn of the century there occurred a geographical shift from a series of harbours used by the dhows to a series of larger harbours now developed as the major ocean terminals. At Mombasa, Kilindini deep-water harbour replaced Old Mombasa harbour, which still retains the form and function acquired in medieval times; Dar es

---

1 *Opposition to the development of Dar es Salaam received a firm setback in 1892 when a cruiser squadron of the German Navy successfully steamed into and out of Dar es Salaam harbour.*

Salaam took the place of Bagamoyo, whilst Tanga replaced Pangani and other minor havens of north-east Tanzania. Fifty years later the same process was enacted when Mtwara was developed as an ocean port in preference to Mikindani and Lindi.

The crystallisation of the traditionally fluid East African port pattern, stimulated by these maritime changes, was linked on the landward side with another technological development – the building of railway lines inland.[1] The intention to build railways – initially as a means of extending and consolidating political control rather than for economic ends – was in fact the immediate cause of the concentration of attention on the mainland harbours, and the building of a small wooden jetty to facilitate the unloading of imported railway construction materials was the first firm mark of confirmation of the selection of these harbours for colonial development.[2] The gradual extension of railway links inland – to Kisumu (1901), Moshi (1911) and Kigoma (1914) – further consolidated the position of the seaports. The subsequent development of the economies of the East African countries has inevitably made great demands upon the seaports; heavy capital investment in both port and railway facilities has increasingly emphasised their vital role, so that today there is a direct relationship between the facilities and traffic of the seaports on the one hand and the extent of the rail feeder system tributary to each terminal on the other. Within the hinterland of Mombasa railway mileage is considerably greater than within other East African hinterlands, and it is there also that most recent rail extensions have been located.

Political influences

The modern East African seaports form part of the transport infrastructure developed to serve the East African countries during their formative period under colonial rule. The emergence of an outline system of seaports and railways was initially deeply affected by the political situation towards the close of the nineteenth century, and subsequently this system has been influenced further by a variety of other political factors. The basic situation created by the initial colonisation and partition of East Africa involved the superimposition of Anglo-German rule and the consequent suppression or incorporation within the new units of the pre-colonial

1 *A summary of the growth of the East African railway system is given in B. S. Hoyle,* 'Recent Changes in the Pattern of East African Railways', Tijdschrift voor Economische en Sociale Geografie, *vol. 54 (1963) 237–42.*
2 *This occurred first at Tanga in 1892, then at Mombasa (1896) and Dar es Salaam (1900).*

political entities both at the coast and inland. The foundation of the modern seaports was closely related to the political division of mainland East Africa into German and British spheres of influence, and the mathematical line of demarcation established in 1886 which today still constitutes, unchanged, the Kenya–Tanzania boundary, exerted a strong influence upon the direction of railway-building from the new coastal ports. The initial selection of port sites and the building of railways inland was influenced profoundly by political circumstances, especially by the need on the part of the colonial powers to establish (and be seen to establish) effective control in their newly acquired dependencies. The division of East Africa between two contrasting colonial regimes created a competitive situation, and the direction of the railways leading inland from Tanga and from Mombasa reflects the Anglo-German race to establish rail communication with Lake Victoria in the 1890s. It was not until 1924 that the two systems were finally linked together. The central Tanganyika line, driven westwards from Dar es Salaam by the Germans between 1905 and 1914, was not connected to the more northerly lines until 1963. Tanga represented a rather unfortunate choice on the part of the Germans as their first venture in port development in East Africa, for apart from the shallowness of the harbour and the failure to establish a direct rail link with Lake Victoria, the subsequent growth of the port has been overshadowed by that of Mombasa and Dar es Salaam, which not only have better water sites but also more extensive rail feeder lines inland.

Because the very temporary political circumstances of the 1886–1919 period in East Africa coincided with the building of the initial arteries of rail communication from the three major mainland terminals, they effectively ensured the continuing division of East Africa into two principal economic compartments, each with its own principal railway and port. After 1919, in an East Africa unified under British administration, little was done to change this situation and it was not until 1948 that the Kenya–Uganda and Tanganyika railways and port services were brought under a single East African administration.[1] As the colonial era progressed, the position of Mombasa as East Africa's principal seaport was increasingly emphasised by the presence within its hinterland of a large European farming community. Stimulated by the British Government, European settlement and development in the highland parts of Kenya augmented port traffic by producing a wide range of export crops and by providing an important market for imported goods. This factor was important, although on a smaller scale, at Tanga and Dar es Salaam; but at all three terminals the physical growth and

1 *The East African Railways and Harbours Administration.*

administration of the ports has been largely controlled by Europeans prior to political independence.[1] The expatriate element in the East African population – both Europeans and Asians – has also long provided a major part of the passenger traffic handled at the seaports.

Thus the colonial relationship, the fundamental political reality in East Africa from the 1880s to the 1960s, was instrumental in the emergence of the modern seaports. Post-independence political changes have also had a considerable effect upon the East African transport pattern as a whole and upon the seaports in particular. At the coast, the incorporation of Zanzibar within the United Republic of Tanzania has done little to stem the decline of Zanzibar port as a trade centre. Originally developed to fulfil a role now performed by the mainland seaport group, Zanzibar has declined into relative insignificance as its hinterland has been progressively reduced by increasingly adverse political circumstances, culminating in the ending of *de jure* Arab rule on the Kenya coast in 1963[2] and of *de facto* Arab control in the offshore islands by the revolution of January 1964. The 1967 East African treaty of co-operation[3] has made provision for the port of Zanzibar to be brought under the control of the East African Railways and Harbours Administration, but such a step is unlikely to have any positive beneficial effect in trading terms, although the efficiency of port working may be improved. Inland, the major political development influencing the continuing growth of the mainland seaports is the problem of finding politically acceptable trade outlets for the landlocked state of Zambia. Since the dissolution of the Central African Federation in 1963, and the subsequent hardening of southern Africa's colour curtain along the middle Zambezi as a result of Rhodesia's unilateral declaration of independence in 1965, Zambia has turned increasingly towards East Africa (especially Tanzania) in both political and economic terms. This is an extremely important development from the standpoint of the Tanzanian seaports, for Zambian copper exports and oil imports have already encouraged a marked acceleration in the rate of physical expansion at Dar es Salaam, and the utilisation of spare capacity at Mtwara is also involved. Evidence of these developments is to be found in the volume of cargo handled at Dar es

---

1   *Tanganyika was the first of the East African countries to achieve political independence, in 1961; Uganda followed in 1962, Kenya and Zanzibar, both in 1963. After the anti-Arab revolution Zanzibar and Tanganyika became the United Republic of Tanzania in 1964.*

2   *See J. W. Robertson,* The Kenya Coastal Strip: Report of the Commissioner, *Cmd. 1585 (London: H.M.S.O., 1961), and also A. Melamid, 'The Kenya Coastal Strip',* Geographical Review, *vol. 53 (1963) 457–9.*

3   *Signed at Kampala, Uganda, on 6 June 1967 by the Presidents of the three East African states. The treaty came into effect on 1 December 1967.*

Salaam – the 1965–6 increase in total cargo movements was 50 per cent – and in the emergence of new or strengthened lines of communication between the two states; initially this is in terms of improved roadlinks and an oil pipeline connecting Dar es Salaam with the Zambian Copperbelt, and potentially in terms of the Tanzania–Zambia railway, now under construction with Chinese aid.

Economic incentives

In parts of the world where economic development is based upon the export of a relatively small range of primary products and the import of manufactured goods, the particularly heavy reliance of the whole economic structure upon overseas trade places singular emphasis upon the vital role of the seaports involved. Although the East African countries are attempting to diversify their economies, particularly by the development of import-substitute industries, overseas trade is still overwhelmingly predominant. In 1965 the total value of East African external trade was £360 million, but trade with African countries outside East Africa accounted for less than £11 million or 0·3 per cent of this total; the separately calculated value of inter-state trade within East Africa was almost £30 million.[1]

Since a very high proportion of the external trade of the East African countries thus passes through the seaports, their vital functional role as a group is quite clear. The emergence of a highly differentiated hierarchy of seaports in East Africa is closely related, however, to the more detailed pattern of economic development in the hinterland areas. Although the complementary importance of forelands in port growth has been emphasised,[2] and although in East Africa many of the initial stimuli involved in port emergence in both medieval and modern times came from the maritime side, the outstanding modern factor involved in the differential pattern of the continuing development of the seaports is undoubtedly the detailed economic geography of the hinterland areas. Where a small number of ports serves a relatively large area, and where the economy of that area is still in a relatively early stage of development, a fairly simple hinterland structure may be expected. Any port has nevertheless many individual commodity hinterlands, which together constitute its composite hinterland. In East Africa the composite hinterland

1 *East African Statistical Department*, Economic and Statistical Review, *vol. 19 (1966)*.
2 G. G. Weigend, '*Some Elements in the Study of Port Geography*', Geographical Review, *vol. 48 (1958) 185–200*. Weigend defines a foreland as '*the land areas which lie on the seaward side of a port, beyond maritime space, and with which the port is connected by ocean carriers*'.

of each terminal consists of a series of fairly well-defined core areas and points of production for export, separated in most cases by wide zones of comparatively little significance in the cash economy. Fig. 13.3 indicates this pattern in a generalised way; the main core areas are shown, and tentative and essentially arbitrary boundaries have been drawn around the composite hinterlands. An obvious feature is that the pattern of core areas within the hinterland of Mombasa is far denser than in the case of any of the other East African ocean terminals, for it contains important core areas for the production of coffee, cotton, sisal and tea together with all the four East African sources of minerals and cement exported in quantity by surface transport.[1] In terms of area there is a strong emphasis on Uganda, but production of cash crops in the highlands of Kenya is intensive and it is mainly from this area that a wide variety of other export commodities is obtained, although none is so outstanding as to compare with those included on the map. The hinterland of Dar es Salaam offers a striking contrast, since it contains only two core areas: the Dar es Salaam–Kilosa sisal zone and the cotton zone in the Lake region around Mwanza. Also, in contrast to the Uganda section of Mombasa's hinterland, the hinterland of Dar es Salaam affords no examples of overlapping core areas. In the case of Tanga the core area of sisal production close to the port is most significant, but the port has some share in the coffee core area of the Moshi–Arusha zone. The hinterland of Mtwara compares in size with that of Tanga, but the cashew-nut core area is rather larger than the sisal core area of Tanga; the contrast in the traffic levels between the two ports reflects the greater weight of the sisal crop and also variations in intensity of production between the two core areas.

No definitive boundary delimits the hinterlands of the East African port group as a whole, but political boundaries set a general outer limit across which very little traffic passes. In the south, however, the extension of the hinterland of the Tanzanian ports (particularly Dar es Salaam) to include the Zambian Copperbelt is a vitally important stimulus to port growth. In the west, traffic with the Congo, Rwanda and Burundi is a significant element in the extra-national hinterlands of both Mombasa and Dar es Salaam. Within East Africa, hinterland overlapping causes some complexity. The hinterland of Mombasa penetrates deeply into Tanzania, and in two areas this results in a degree of overlap with Tanzanian ports: in the case of coffee exports from the Bukoba region and of cotton shipments from Musoma, the deciding factor is the greater convenience of movement

1 *Diamonds are exported from Tanzania by air; some cement is exported from Dar es Salaam.*

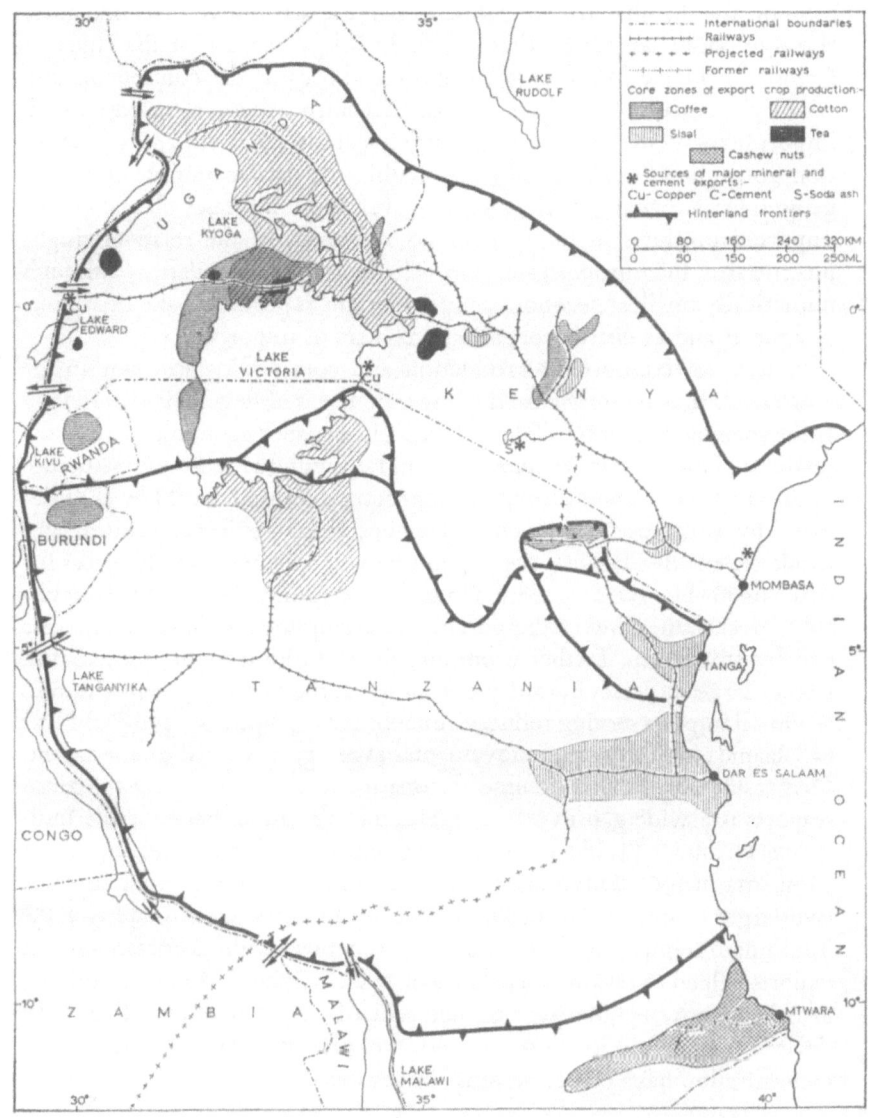

Fig. 13.3   Some economic factors involved in the growth of East African seaports

I

by railway and lake steamer to Mombasa than to Dar es Salaam; in the case of the Moshi–Arusha coffee zone, exports move predominantly through Mombasa rather than through Tanga because of the superior facilities available in the larger seaport and because the commercial and financial control of the coffee trade is centred there. The pattern of import demand in the hinterlands largely parallels that of export production, for in broad terms production within the cash economy emphasises geographical variations in income levels and increases demands for imported products in the more favoured areas. In the relatively high-income areas the European and Asian elements in the population, although numerically small, are economically important as producers and organisers of exports and as distributors and purchasers of imports.

Patterns of commodity production and commodity flow within the hinterlands thus re-emphasise the disparate levels of development attained at the mainland seaports of East Africa. In broader economic terms, the fortunes of the seaports are closely bound up with the economic structure of the areas served, and through that structure with the world level of demand for East Africa's main commercial products – coffee and cotton, and to a lesser extent sisal, tea, copper and cloves. Declining world prices for sisal, and problems associated with world overproduction of coffee, inevitably affect traffic flow at the ports by lowering the rate of traffic increase and by stimulating further economic diversification in the hinterlands. Recent large increases in tonnages handled at the ports are mainly due to crude-oil imports serving refineries established at Mombasa (1963) and Dar es Salaam (1966); dry cargo movements have not grown to the same extent. One result of all the economic factors at present affecting East African seaports is a wide gap in traffic levels and in facilities between the individual terminals (Table 13.3). The introduction of crude-oil imports at Mombasa in 1963 temporarily widened still further the gap between the two larger terminals, but at Dar es Salaam imports of crude oil (for the Tanzanian refinery and for Zambia) together with Zambian copper exports helped to restore the balance in 1966. Together Mombasa and Dar es Salaam now handle over 94 per cent of the total traffic of the East African mainland port group, and the pressures resulting from this concentration have caused serious congestion.

Conclusions

The emergence of a group of seaports in a developing economy involves the interplay of a wide range of factors. Many of these factors are associated directly or indirectly with the physical environment, others with economic or political conditions or with technological changes. At any given point in time the range of variables is complex; some of the factors involved are not themselves geographical or economic, but nevertheless have economic and spatial effects. Some factors may operate only on a very local scale, such as those associated with port sites; others, such

Table 13.3

Percentage of total cargo traffic at East African mainland seaports handled at individual terminals, 1956–68

Year	Mombasa	Dar es Salaam	Tanga	Mtwara*
1956	68·3	21·4	7·1	3·2
1957	70·4	20·1	6·1	3·4
1958	71·6	20·1	5·2	3·1
1959	70·9	20·4	5·5	3·1
1960	69·8	21·9	5·2	3·1
1961	71·8	19·3	5·9	3·0
1962	71·6	20·0	5·4	3·0
1963	73·9	18·6	4·8	2·7
1964	77·3	16·0	4·2	2·5
1965	78·2	16·0	3·6	2·2
1966	74·1	20·5	3·4	2·0
1967	68·8	25·9	3·4	2·0
1968	68·3	26·5	3·0	2·2

* *Includes Lindi 1956–62.*

Source: *Calculated from East African Railways and Harbours,* Annual Reports 1956 *ff.*

as the distribution of rainfall or of railways, represent conditions within the hinterland areas served by the ports; a third group, on a continental or world scale, indicates a reaction to far wider circumstances and events. Some examples of each of these types of factor are given in Table 13.4; each category and each factor is subject to time changes, so that a third dimension is introduced and the framework suggested by the table can be moved backwards or forwards in time with contrasting effects. The fundamental fact which this discussion underlines is that seaports do not

Table 13.4

Some examples of factors involved in the emergence of the modern East African seaport group

Scale	Physical	Historical	Technological	Political	Economic
Local	Site conditions: East African coastal geomorphology	Earlier port hierarchies	Increasing size and manoeuvrability of shipping	Arab coastal strip	Availability of exports; port industries, labour
Hinterland East Africa	Incidence of rainfall and highland areas; varied crop environments	Indian Ocean trade circulations; interior exploitation; slaves and ivory	Development of railway system	Anglo-German penetration; colonial era	Establishment of cash economy
Continental World	African barriers to earlier penetration	Isolation of Africa; Islamic cultural history	Suez Canal; industrial growth in Europe	Africa; penetration, colonisation, exploitation, development, decolonisation	World demand and price levels

grow in isolation but in response to changing opportunities and demands at different times, over different areas, on different scales and with different intensities. Whilst this applies to seaports anywhere in the world, a point worth re-emphasising is that in developing areas many of the stimuli involved in the establishment and growth of modern seaports have come from outside the areas concerned, and have come by sea; thus in the case of East Africa the emergence of the modern seaports since the 1890s is a continuation of a tradition going back two thousand years. Only in the twentieth century, as the economic development of East Africa has proceeded, have factors from the landward rather than the maritime side become the predominant influences.

An appreciation of the dynamic element in port geography is important at the present time, for the modern seaports of East Africa are still growing and changing. The East African port hierarchy of 1980 or 2000 will not be the same as that of 1970, itself different from the pattern of 1950. Mombasa is likely to consolidate still further its predominant position, since it possesses unrivalled advantages in several respects. Dar es Salaam is already responding positively to the challenge of Zambian traffic, and Mtwara may also do so and thereby be rescued from commercial stagnation. Tanga is not likely to see rapid development in the immediate future, because of site and situation problems and falling sisal prices, but in the long term the possibility of growth at a higher level should not be entirely ruled out. Zanzibar, after its brief period of glory arising from the very unusual political and economic circumstances of the mid-nineteenth century, now seems destined to play the role of a minor local port. Adjustments are thus taking place within the modern hierarchy as factors comparable to those that have influenced its emergence continue to operate. Increasing concentration on the two main terminals reflects the combined effect of these factors, past and present, and because of them it is at Mombasa and Dar es Salaam that further new facilities and new cargo-handling methods are now being introduced.

14. Ports and Economic Integration in Madagascar

William A. Hance

William A. Hance is Professor of Geography and Chairman of the Department of Geography at Columbia University, New York. His numerous publications on Africa include The Geography of Modern Africa (1964) and African Economic Development, 2nd (rev.) ed. (1967).

The port situation of Madagascar is characterised by a large number of poorly equipped lighterage installations and three ports with deep-water facilities, only one of which is modern and adequately equipped. The dispersion of ports along both coasts both reflects and contributes to the poorly integrated transport and economic structure of the island, which is, in turn, a major explanation for the relative stagnation of the Malagasy economy over at least the past three decades.

In assessing the validity of this thesis, this chapter will (1) give evidence of the stagnation; (2) describe the port and transport complex of the Republic; (3) examine the transport infrastructure in the context of the island's economy with particular reference to the geographical sectionalism of its producing and consuming 'islands'; (4) expose other possible explanations for the relatively retarded economic growth; and (5) present a final assessment of the hypothesis with some indications as to how the port and transport position might be improved to achieve a more satisfactory integration of the island's economy and thus reduce the impact of what is seen as a major retarding influence on economic growth.

Stagnation of the Malagasy economy

Some of the best measures of the relative stagnation of the island's economy are revealed in the trade patterns of the past three decades. In the period 1938–65 exports increased by value 3·8-fold as compared with a 12·2-fold growth for tropical Africa; imports grew 8·1 times as compared with 11·3 times for tropical Africa. The average volume of exports in the five-year period 1963–7 was only 36·7 per cent above that of 1938. Of the top twenty-two exports in 1967, including 90 per cent of total exports by value and all commodities with a value over 0·5 per cent of the total (Table 14.1), only five showed any marked achievement in production levels over 1938. Rice, sisal and bananas among the five have shown wide fluctuation in volume and value over recent years, with imports of rice sometimes exceeding both the volume and value of exports.

Madagascar shipments in the five-year period to 1967 of its main export, robusta coffee, which accounts for about 32 per cent of the total value of exports, averaged only 10 per cent above the 1938 tonnage. The island failed to participate in the coffee boom of the 1950s enjoyed by many African countries, its share of African coffee exports declining from 9·6 per cent in 1947–9 to 5·0 per cent in 1964–6.

Products which did not appear on the 1938 list of exports accounted for only 6 per cent of the total value of exports in 1967, while as many commodities disappeared from the list as appeared on it. The unusual variety which has characterised Malagasy exports has brought some stability despite wide fluctuations in both the volume and value of many individual exports, but it does not appear to have stimulated growth significantly, and the strength of the Malagasy economy compares unfavourably with many of its one-product-oriented neighbours on the continent. Finally, there has been a markedly unfavourable balance of trade since 1947. Since transit trade, tourism and other invisible items are of minor significance, Madagascar's dependence on foreign assistance to achieve an overall balance of payments remains very heavy.

Many other measures may be cited to confirm the relatively poor growth record of the island's economy. The estimated per capita gross domestic product is below that of the Central African Republic and ranks eighth among fourteen francophone African countries. Wage employees totalled under 200,000 in 1965 of the total population of 6,335,810 and there was marked stagnation, if not retrogression, in non-agricultural employment in the period 1960–7.

The record of growth in the several productive sectors has been unsatisfactory. Agriculture, which is the dominant activity, has not advanced well, with numerous plans for improvements failing to meet their goals, sometimes by a very wide margin. The island could, because of the absence of the tsetse fly, develop a mixed agriculture, but there is now almost complete separation of livestock grazing and crop farming, and the livestock population, including an estimated 8 million cattle, contributes only meagrely to the national income (about 6 per cent) and to the exports (about 9 per cent) of the country.

Mining is only of minor significance to Madagascar. The two main

Table 14.1

Madagascar exports by value and percentage of total value, for selected years, 1938–67

Commodity:	Exports by value (million local francs)					Exports by % Total Value				
	1938	1955	1965	1966	1967	1938	1955	1965	1966	1967
Coffee	261	6,193	7,133	7,593	8,122	31·9	43·4	31·5	31·5	31·6
Sugar	—	n.a.	1,162	1,485	2,138	—	n.a.	5·1	6·2	8·3
Rice	16	1,226	676	1,052	1,854	2·0	8·6	3·0	4·4	7·2
Vanilla	75	713	2,437	2,216	1,672	9·2	5·0	10·8	9·2	6·5
Live animals	n.a.	n.a.	1,295	1,318	1,040	n.a.	n.a.	5·7	5·5	4·0
Petrol products	—	—	—	166	983	—	—	—	0·7	3·8
Raffia	29	503	682	812	918	3·5	3·5	3·0	3·4	3·6
Clove products	50	637	744	579	876	6·1	4·5	3·3	2·4	3·4
Meat conserves	n.a.	395	721	721	780	n.a.	2·8	3·2	3·0	3·0
Sisal	5	345	1,340	1,006	743	0·6	2·4	5·9	4·2	2·9
Tobacco	7	957	1,037	1,267	711	0·9	6·7	4·6	5·3	2·8
Essential oils	n.a.	n.a.	473	594	555	n.a.	n.a.	2·1	2·5	2·2
Beans and peas	21	310	848	708	537	2·6	2·2	3·7	2·9	2·1
Graphite	18	465	462	455	432	2·2	3·3	2·0	1·9	1·7
Hides and skins	43	225	351	586	372	5·3	1·6	1·6	2·4	1·4
Peanuts	n.a.	305	394	295	364	n.a.	2·1	1·7	1·2	1·4
Bananas	—	—	221	403	264	—	—	1·0	1·7	1·0
Mica	29	127	241	258	260	3·5	0·9	1·1	1·1	1·0
Pepper	1	212	317	260	259	0·1	1·5	1·4	1·1	1·0
Tapioca	17	218	211	246	240	2·1	1·5	0·9	1·0	1·0
Vegetable oils and cake	n.a.	n.a.	n.a.	152	153	n.a.	n.a.	n.a.	0·6	0·9
Manioc	29	80	127	205	152	3·5	0·6	0·6	0·8	0·6
Total above	601	12,911	20,872	22,377	23,425	73·4	90·6	92·2	92·7	91·0
Total all exports	819	14,268	22,632	24,132	25,711					
Percentage exports to imports	136	67	66	69	72					

Totals and percentages may not agree owing to rounding.

Source: Government of Madagascar, Bulletin Mensuel de Statistique.

minerals have long been graphite and mica, which accounted for only 2·7 per cent of exports by value in 1967 as compared to 5·7 per cent in 1938. A variety of other metallic and non-metallic minerals are produced from small and scattered deposits, but only chromite promises to add significantly to mineral exports. It will be produced near Andriamena, transported 88. km (55 miles) by road to a short rail extension from the Lac Alaotra branch, and railed to Tamatave for export. From mid-1969 planned annual output of 51 per cent concentrate is 100,000 tons A small bauxite deposit is under investigation in the south, while prospecting for oil is scheduled in offshore areas of the Mozambique Channel. Earlier searching along the western side of the island revealed no commercial occurrences.

Madagascar has a relatively limited list of modern manufacturing establishments, although its craft production is well developed, particularly on the central highlands. Most plants are small and there is considerable unused capacity in numerous branches. The more important modern industries include textiles and clothing, shoes, cement, tobacco products, matches, beer, and a 600,000-ton petroleum refinery which went on stream in 1966 and which also serves the Comoros Islands and Réunion.

The position with respect to energy production and consumption compares unfavourably with African countries and reflects the retarded economic growth of the island. Production increased 5·5 per cent per annum in the period 1952–62 and 7·0 per cent from 1962 to 1966; these rates are below the African average, and per capita consumption of electricity is only about 11 per cent of the average for the continent. On the favourable side, all agglomerations over 5,000 are now electrified, although the percentage of inhabitants served in them usually remains quite low. Most stations are quite small, the largest being the 22,500-kW. Mandraka hydro-electric plant 69 km. (43 miles) from Tananarive. There is considerable unused capacity and electricity is generally expensive.

The economic growth of Madagascar, then, appears to be far from satisfactory and to compare unfavourably with other African countries both in recent years and over the longer-run period since before the Second World War. The dependence on foreign-managed operations continues to be important in the production of sugar, rice, sisal, bananas and other crops, minerals, and manufactured goods, which means that the Malgaches have made an even less impressive record than might otherwise be concluded.

A somewhat alarming component is added when one enters the demographic factor into the equation. Thanks to greatly improved health facilities, and particularly to a massive anti-malaria campaign, the excess of births over deaths increased from an estimated 10,978 in 1946 to 184,000 in 1961 and 310,000 in 1965. The estimated rate of population increase is now about 3·5 per cent per annum; not many production indexes are sufficiently high to provide any respectable improvement in the face of such a rapid rate of population growth. There are marked variations of population density, and according to the Commissariat Général au Plan the demographic increase has created zones of population pressure where very low standards of living now pose serious problems. Finally, the individual farmer does not appear to have adjusted his techniques or output to the now rapidly rising population on the island.

The transport complex of Madagascar

Madagascar has a rather scanty road system, 864 km. (540 miles) of railways in two lines, an unusual canal along the east coast, a remarkable air network, and more than sixteen ports, seven of which handle more international than coastal traffic.

Ports

Malagasy ports handled 869,500 tons of international traffic in 1965, a rather feeble tonnage considering the size of the island and the length of its coastline. The leading port, Tamatave, handled 54·4 per cent of the international cargo and 43·7 per cent of total port traffic. The top six ports together handled 83·3 per cent of total port traffic in 1965. The major explanations for the relatively low tonnage handled in Malagasy ports are the underdevelopment of the economy and the character of the exports. Many export commodities have high value in relation to their bulk (coffee, vanilla, clove products, tobacco, essential oils, pepper, etc.), and there are no bulk mineral exports such as have justified the expansion of many African ports.

Only Tamatave can be considered a modern port, though its facilities are not adequate for the present tonnages. Tamatave, Diégo-Suarez and Tuléar are the only ports that have some kind of deep-water quay; many ports are no more than poor lighterage harbours with no protection for ships at anchor and with minimal shore and floating installations. The east coast is noted for the absence of indentation (Antongil Bay is open to the prevailing winds); the west coast has other physical difficulties, particularly the silting of river mouths. Ironically, Madagascar has one of the finest harbours in the world at Diégo-Suarez, but its value is greatly reduced by its location at the northern tip of the island and by its restricted hinterland.

Tamatave is the only complete and modern merchant port, and more than any other is a gateway for the island, not just for a region. In 1965 it handled 64·0 per cent of Malagasy imports, 35·6 per cent of its exports and 23·4 per cent of coastal traffic. This pattern reflects the importance of Tamatave as the gateway for the leading consuming area on the island – the highlands centred on Tananarive – and the use of Tamatave as one of the four main transhipment points for coastal traffic (Fig. 14.1).

The harbour of Tamatave is formed by a series of coral reefs plus a dike which give protection to the semi-enclosed deep-water area. Extension of the present dike would be required to give full protection to the harbour. The port has two deep-water moles, but only one berth with alongside depths of 10 m. (36½ ft), plus lighterage and service quays. It is fairly well mechanised with twenty-nine quay cranes, eleven mobile cranes and one 120-ton floating crane, as well as numerous pieces of smaller mechanical equipment.

Majunga, on the west coast, is Madagascar's second-ranking port with a total traffic of 245,300 tons in 1965, about equally divided between overseas and coastal traffic. It has a well-protected harbour but no deep-water quays. Diégo-Suarez, with one commercial deep-water quay (it is also a naval base), ranks third among Malagasy ports. Tuléar, the chief port for the south-west, has a safe anchorage and a deep-water berth of sorts in the form of a 1,280-m. (1,400-yd) jetty ending in a small pier with alongside depths of 8·5 m. (28 ft).

The dispersion of activity among sixteen ports means that available funds are stretched too far to permit adequate installations anywhere. Nine of the sixteen handle less than 50,000 tons annually, and fourteen handle less than 100,000 tons. Since a yearly rate of about 200,000 tons is usually needed to justify deep-water facilities, it is obvious that Madagascar cannot expect to see the construction of modern facilities in more than a few locations.

Railways

Madagascar has two metre-gauge railway lines: the Tananarive–Côte Est Railway (T.C.E.), connecting the capital with Tamatave and having branches to Antsirabe and Lac Alaotra, and the Fianarantsoa–Côte Est Railway (F.C.E.), running from Manakara, an open-roadstead port, to the capital of the Betsiléo country in the highlands (Fig. 14.1). Both lines have very difficult profiles, with maximum gradients of 2·5 and 3·5 per cent, minimum curve radii of 50 and 80 m. (54·7 and 87·5 yds), numerous tunnels, bridges and culverts and much cutting and filling.

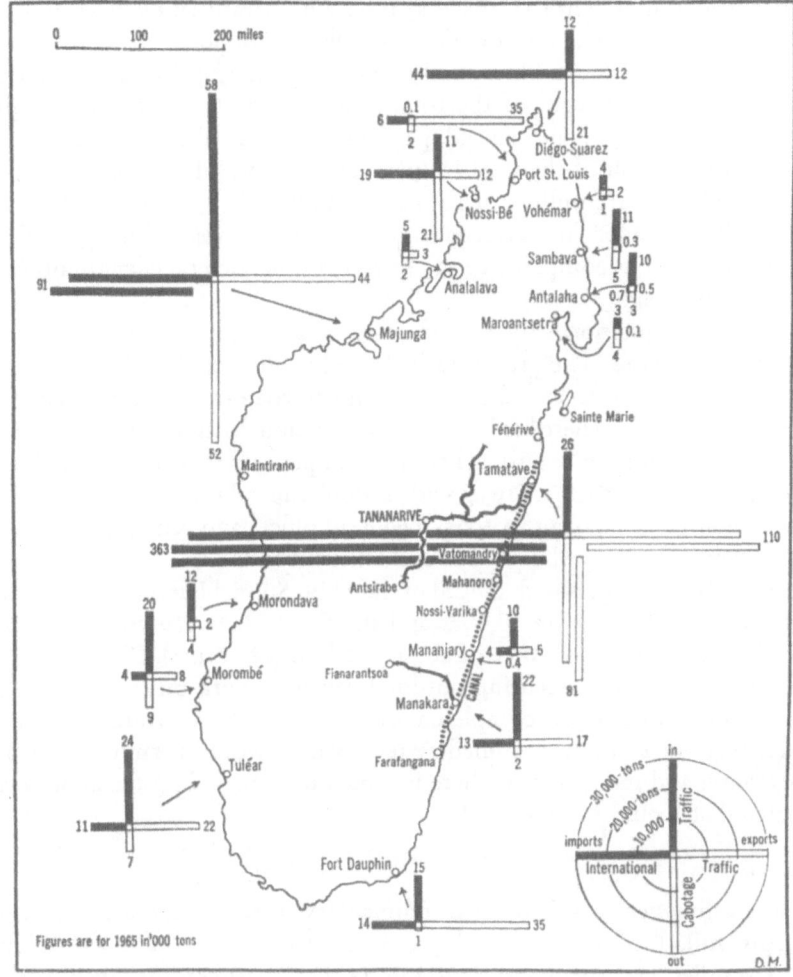

Fig. 14.1 *International and coastwise traffic at the ports of Madagascar in 1965. Inward movements are shown in black, outward movements in white; international movements are shown by horizontal bars, coastal traffic by vertical bars.*

The lines are now completely dieselised, the rolling stock totalling forty-seven locomotives, eighteen diesel railcars, ninety-eight passenger cars and 1,014 freight cars, in 1966. The total tonnage carried on the two Malagasy lines was 604,000 in 1966, which compares with 619,000 tons in 1955. The freight-ton-mileage, however, increased from 88 million in

1955 to 104 million in 1966. In 1965 the traffic, measured by share of total freight-ton-mileage, was divided as follows: imports 53·9 per cent, exports 3·2 per cent and local traffic 42·9 per cent. The T.C.E. line carries more than 90 per cent of the total freight traffic. Direction of freight movement is highly unbalanced, particularly on the T.C.E., with up-line traffic 81 per cent of the total in 1965. The number of passengers carried increased from 1·1 million in 1938 to 2·3 million in 1955 but declined to 2·1 million in 1966; the average length of journey was 70 km. (44 miles) in the last year as compared to 317 km. (173 miles) average length of haul per ton of freight.

Total receipts normally exceed costs of operation but do not cover renewal expenses. The precarious balance has also required high average rates, which helps to explain a recent shift to road transport, particularly for export traffic. There has been successful rationalisation and modernisation of operations since 1950. There was a 21 per cent reduction in numbers employed from 1938 to 1954 with a doubling of traffic, and a further reduction of 12 per cent to 1965. The head offices and central workshops at Tananarive are well equipped and efficiently run.

A study was begun with the aid of the U.N. Special Fund in 1965 on the feasibility of linking the T.C.E. and the F.C.E. between Antsirabe and Fianarantsoa over a distance of about 290 km. (180 miles). If the government is successful in securing outside financing, work on the line might begin in 1970 and be completed by 1974. While it would have the advantage of forming a unified system, which would permit savings in operation and maintenance, there is some doubt regarding the additional traffic that would be generated.

Roads

The existing network of roads is rudimentary: there are about 15,145 km. (9,465 miles) of national and provincial roads, of which 2,320 km. (1,450 miles) are paved and 7,140 km. (4,460 miles) are improved to all-weather standards (Fig. 14.2). An additional 12,385 km. (7,740 miles) of tracks are of doubtful negotiability. The pattern consists of somewhat distorted backbone running the length of the island, with occasional ribs extending outwards to the more important ports. Even parts of the dorsal road are normally in poor condition. Road movement is also inhibited by wash-outs, disrupted ferry services, limited bridge and ferry capacities, and inadequate maintenance over large stretches. Many coastal sub-regions are very poorly connected with the rest of the island and are therefore primarily dependent on coastal vessels for their national and international

Fig. 14.2   The road system of Madagascar

shipments. The best road system radiates from Tananarive, which is a kind of hub most of whose spokes do not extend to the rim. That city is connected by road with the two main ports, Tamatave and Majunga.

Inland waterways

The most interesting inland waterway is the Pangalanes Canal, which runs for about 640 km. (400 miles) along the east coast from Foulpointe to Farafangana (Fig. 14.1). It connects a series of lagoons by cutting across the sills or *pangalanes* that separate them. Only about a quarter of the canal can take 30-ton barges; much of it is open only to canoes.

To some extent it is regrettable that the canal exists, because it has been tempting to try to enlarge it and to consider it as a potential link to Tamatave for all the regions it serves. This has involved considerable expenditures from time to time, but the results have been disappointing in increased traffic, have not eliminated the need for local road transport, and have required several otherwise unnecessary transhipments. It would have been better to construct an all-weather road along the coast, which would have permitted the closing of several inadequate ports and the concentration of their traffic on Tamatave.

The rivers of Madagascar are of only local importance in the carriage of goods, with the exception of the Betsiboka, which is fairly heavily utilised for about 80 km. (50 miles).

Air transport

Madagascar has a surprisingly well-developed air transportation complex, which reduces the isolation of many points, at least for movements of people. With more than 199 airfields and landing strips, of which 57 are served by scheduled flights, it has the densest airline net in francophone Africa (Fig. 14.2). Air Madagascar (formerly Madair) exploits the national routes, which have shown little growth in traffic since 1960; Air France provides technical assistance to this line plus exterior service, which increased by 122 per cent from 1960 to 1965. The main airport for international flights has shifted from Arivonimamo, some 80 km. (50 miles) from the capital, to nearby Ivato, which has long been the main internal field.

Economic sectionalism on the island

The economy of Madagascar, as has been suggested, is very poorly integrated from the geographical standpoint. The inadequate overland transport net and the scattering of commercially important regions, particularly the areas producing for the export market, create a centrifugal effect which has serious impacts on the developing economy. The large number of ports and the insular pattern of development reflect and enforce the lack of integration.

The export 'islands' are mainly peripheral in location (Fig. 14.3), including the scattered east-coast nodes producing coffee, clover and vanilla, the sisal estates in the Mandrare valley tributary to Fort Dauphin, the sugar, pepper, essential oils and vanilla-producing zones of the Sambirano and north Mahavavy valleys and of Nossi-Bé in the north-west, and the isolated 'islands' along the west coast shipping peas, sisal, cotton and livestock products. Ports which serve these and other discrete 'islands' handled 44·5 per cent of international exports in 1965 and 33·2 per cent of outgoing coastal movements. The fact that they handled only 16·9 per cent of imports reflects a serious imbalance between their development and that of the main consuming regions, while their accounting for 56·1 per cent of incoming coastal movements suggests the high cost of supplying these isolated nodes.

Only a few of the ports have hinterlands of some size, namely Tamatave, Majunga, Manakara, Tuléar and Fort Dauphin. Others, such as Diégo-Suarez and Nossi-Bé, benefit from coastal shipments though they do not have extensive hinterlands. The remaining ports serve only restricted regions and some would certainly not exist if better land communications were available.

Heightening the disequilibrium of the island's economy is the location of the main consuming region in the central highlands focused on Tananarive, but stretching from north of that city southward to Fianarantsoa. This is evidenced by the concentration of political, cultural, commercial, construction and manufacturing activities at the capital, whose metropolitan population was 321,654 in 1965. A rough measure of this position is seen in the high percentage of international imports landed at Tamatave (64·6 per cent), of which 40 per cent moves up-line on the T.C.E. Other measures include the very high proportion of total electricity output consumed in the Tananarive area and the registration in 1965 of over half of the 62,016 motor vehicles on the island in that province.

The high cost of transport resulting from the lack of integration on the island is suggested by tracing the movement of individual shipments. Exports from one of the isolated nodes will usually be transported by road to the port, lightered to vessels standing offshore or moved by coastal vessel to the nearest main port (coastal traffic exceeds international

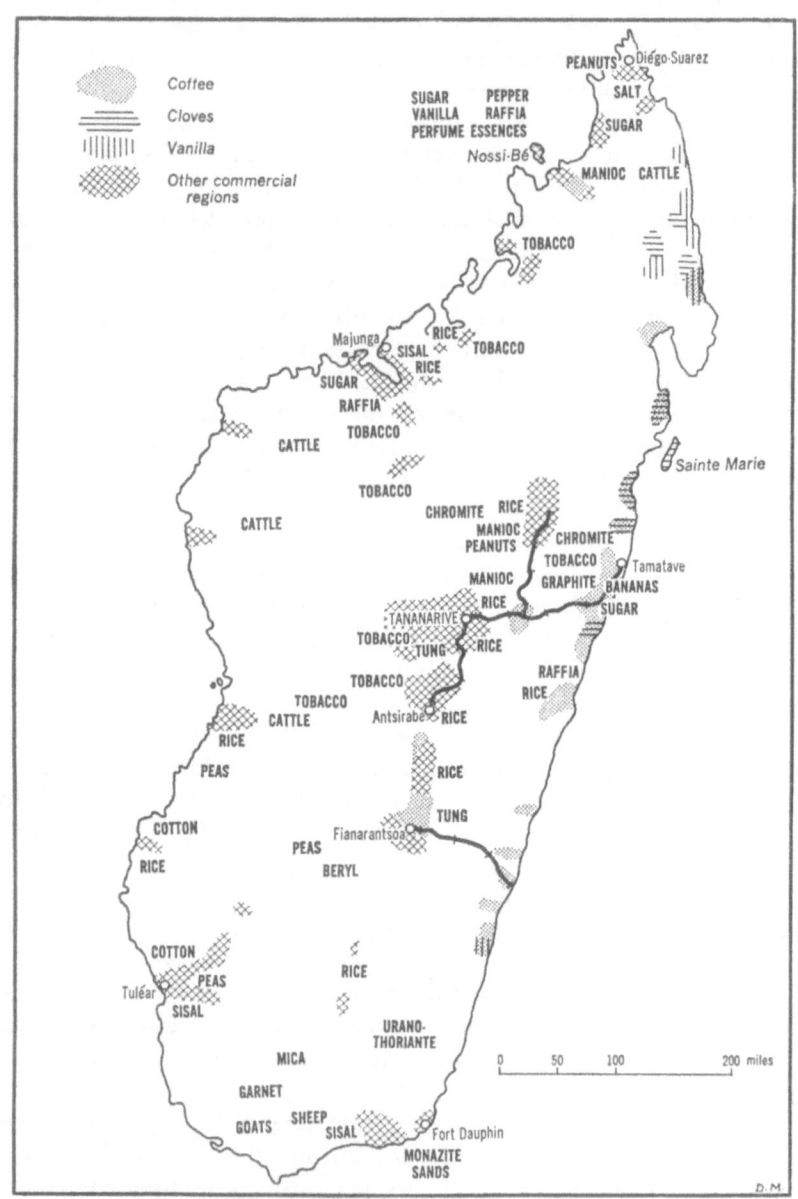

Fig. 14.3 *The commercially productive regions of Madagascar*

shipments in nine ports, and total coastal tonnage in all ports is 52·4 per cent of total international shipments), and transhipped at the port of embarkation.

The most inhibiting situation arises with regard to industry catering for the domestic market. The largest segment of such industry is located in the Tananarive–Antsirabe area. If a given plant consumes raw materials from abroad, these must first be landed at Tamatave and then railed to the plant; finished goods destined, say, for an isolated east-coast node must then be railed down-line to Tamatave, sent by coastal vessel to the port serving that node and then moved by road to the final destination. This often makes for a much more costly product than could be delivered from Europe. The difficulty of supplying the highly fragmented market of Madagascar is a major explanation for the relatively restricted list of industries thus far established on the island.

The disintegrated character of the Malgache industrial market contrasts sharply with many African countries, whose main manufacturing centres usually have the best connection of any city with the rest of the country and where the focusing of rail and road systems is much more marked. The major representation of market-oriented industries is often at the seaboard, as at Dakar, Abidjan, Accra, Douala and Dar es Salaam, or at the inland end of an axis still well placed to serve the entire national market, as in the cases of Kinshasa, Brazzaville and Nairobi. Only Mozambique and to a lesser extent Angola are somewhat comparable to Madagascar; each has a string of ports along the coast serving rather discrete domestic hinterlands. Both Portuguese territories, however, have two advantages not shared by Madagascar – the presence of numerous excellent harbours and the existence of rich extra-national hinterlands which provide valuable transit trade.

While it is true that Madagascar could not exist in its present stage of development without numerous outlets to the sea, and that the country cannot afford a fully integrated transport pattern, it is obvious that the markedly sectionalised character of the economy is a major deterrent to development. One might almost conclude that Madagascar's island character is a disadvantage; indeed, economically it functions much more like an archipelago of islands than like an island continent, which it has sometimes been called.

Other explanations for Malagasy stagnation

Madagascar, it will be recalled, has seen relative economic stagnation not only in recent years but extending over at least three decades. We have seen that the dispersed and isolated nature of production and consumption 'islands' provides one major explanation for this atypical pattern. In this section other possible explanations will be examined to secure at least a subjective idea of their comparative importance.

Physical limitations

There are many shortcomings in the physical endowment of Madagascar. Gautier wrote that it had the colour and fertility of a brick. A large area is plagued by severe soil erosion. The largest portion of the better-watered areas, including the region inhabited by the most advanced segment of the population, has a difficult topography, the vast bulk being in slope. Most of the west and south is handicapped by meagre and uncertain precipitation. Tropical storms occurring with irregular but certain fury often cause severe damage, especially on the east coast. And problems of access are often great.

But none of these, nor all of them combined, can provide an adequate explanation. First, because the island also has substantial physical assets, including a range of climate types permitting great variety in production; some good to excellent alluvial and volcanic soils; an enormous area capable of being cropped, while at present only about 3 per cent is so utilised; and opportunities to develop irrigation in the western river basins. Second, because its deficiencies are not unique to Madagascar among African countries and other countries have made satisfactory progress despite them. And Madagascar has one asset not shared by most African countries – absence of the tsetse fly – which permits development of an integrated crop–livestock agriculture and a greater interest in livestock, advantages which have not been realised upon satisfactorily.

Another factor frequently cited to explain Malagasy difficulties is its remoteness from West European and other markets. This seems to the present writer to be a convenient rationalisation. It is true that Madagascar lies somewhat off the beaten track, but it is served by several liner-freight services. If isolation were really as severe as it is said to be, Mauritius, which is several days further away on the same services, should have equal problems. In fact it exports about three-quarters as much as Madagascar by value, from an area about one-three-hundredth as large.

Economic explanations

A number of economic considerations contribute to Malagasy development problems, but none of them appears adequate to explain the long-run nature of the retardation. Lowered prices for certain exports, which have adversely affected the export earning of many African countries in the past decade, have also influenced Madagascar, but these cannot relate to the thirty-year span going back to pre-war years.

Madagascar does have an excessive hierarchy of middlemen, with many large companies in foreign, largely French, hands and much petty trade conducted by Indians, Chinese and Réunionnais, but this pattern does not differ radically from numerous African countries which have made more notable economic progress. The existence of some share-cropping, under which most of the tobacco and about 15 per cent of the rice is produced, inhibits initiative, but it is scarcely of major importance.

More significant has been the imbalance between rural and urban development. There has probably been an excessive concentration of available resources on Tananarive and a disproportionate allocation of government funds to the new, privileged government employees concentrated there. President Tsiranana himself has denounced the excessive proliferation of civil servants, whose Malagasy component has increased from 14,500 in 1947 to 25,150 in 1960 and 50,000 in 1967. The number of French officials declined from 4,000 in 1953 to 1,736 in 1964. The proliferation of government employees helps to explain the high percentage of the total budget absorbed by ordinary expenditures and the relatively low percentage available for the development budget. These conditions are, however, common among African countries, as has been well noted by René Dumont in his *False Start in Africa*.

It cannot be denied, however, that there has been a failure to enlist the initiative of rural residents in efforts to develop the economy. It has been estimated, for example, that fewer than 10 per cent of the peasant families on Madagascar are producing cash crops. As a consequence of both inadequate agricultural advisory work and apathy on the part of the rural population, there has been little change from the characteristic pattern of agriculture in the central highlands, which may be described briefly as rice in the valleys and cattle on the hills.

Turning to another economic factor, there has been frequent criticism of the shipping companies serving Madagascar, which are accused of excessively high rates, seen as a major deterrent to economic development. I know of no island, however, which does not attribute much of its problems to the high rates and inadequate service of its shipping links. The companies involved do not deny that high rates prevail, but state that they reflect the deficiency of Malagasy ports, the small tonnages involved and the imbalance between imports and exports. Two domestic shipping companies have been formed since 1960, both of which are aided by one or more of the conference members. There are now three ocean-going freighters in the fleet plus numerous coastal vessels, but, as has been the case with the formation of other African national lines, no reduction in freight rates has occurred. Closure of the Suez Canal has lengthened the trip between Madagascar and Europe from fourteen to twenty days and brought increased rates, but the island is less at a disadvantage in this respect than East Africa, Somalia and the Sudan. Consideration is now being given to the introduction of container services to reduce breakages, pilferage and insurance charges.

The inbred character of franc-zone economies may also be cited as an explanation for Malagasy problems. The high dependence over many years on France for assured markets at subsidised prices, and the reciprocal protection for French goods in Malagasy markets, has inflated costs and reduced the competitiveness of island products on the world market. The adjustments required by the association with the E.E.C. may have a desirable cathartic effect, but they are also likely to be quite painful.

One final possible economic explanation for the retardation as measured by the slow growth in trade may be the relatively well-developed exchange economy that has existed for some decades on the central highlands. The range of locally produced foods, building materials, household furnishing and craft items is unusual by tropical African standards, the Zoma or market at Tananarive being most impressive from this standpoint. None the less, as has been noted, the central highlands remain the major consuming area for imported goods.

Cultural–political explanations

A major explanation for the long-term relative stagnation of the Madagascar economy, and one which is probably even more important than the lack of integration of the island's economy, is the inhibiting character of certain features of Malagasy culture. There is an exceptional insularity among the people of Madagascar, a conspicuous ethnocentricity. They do not feel themselves African or wish to be considered African. Indeed, at least the Merina and Betsiléo are not, their ancestors having migrated from Indonesia. A high respect for custom and the continuing feudalism of the Merina, whose caste divisions are based on birth and wealth, are not conducive to change. But most important is the so-called 'cult of the dead', which embodies a profound fatalism and a concentration upon death, the family tomb being the most sacred of all hallowed places. It involves a strong desire to emulate and to live like one's forebears, for not to do so is to transgress taboos and to invite the vengeance of the spirits.

This 'cult' helps to explain the noted docility of the Malagasy people, their shunning of innovation, their passive resistance to economic progress. The numerous ceremonies involved even have direct effects on the economy by cutting the number of workdays and by involving the sacrifice of some of the best cattle. These aspects of Malagasy culture, then, add up to one of the most important elements inhibiting dynamism and economic advance.

Several political factors may also be given in partial explanation for the economic retardation of Madagascar. The 1947–8 revolt, which reflected Malagasy resentment of repressive French policies, left a partial vacuum which persisted for at least seven years. There was particular aversion to the Réunionnais, of whom there were some 23,960 in 1954 and who made up a large portion of the planter and small merchant class. Comprising a more permanent and less educated element than the metropolitan French on the island, they were accused of abusing the labour requisition system, underpaying their employees and engaging in usury, and they were in turn the main 'European' victims of the revolt. This revolt affected Madagascar during the period when most African countries were making striking progress in economic development. There has also been some resentment of the Indians and Chinese, of whom there were 14,604 and 8,045 in 1965.

Despite having a common language and many common cultural elements there continues to be a lack of unity among Malagasy tribes, which have never been welded into a cohesive nation. The Merina consider themselves superior and resent control of the government by the littoral tribes, collectively called the *côtiers*. The *côtiers*, on the other hand, resent the Merina, who occupy a disproportionate number of administrative posts throughout the island. In 1947, for example, Merina held all but 2,000 of the 14,501 civil service posts occupied by indigenous people; in 1962 they still accounted for 59 per cent of all Malagasy functionaries, though they comprised only 23 per cent of the population.

At the time of independence in 1960 there was a *crise d'autorité*. Peasants confused independence with anarchy, refused to obey the local officials and resisted the payment of taxes. There were also numerous cases of graft. To some extent these difficulties, which were not peculiar to Madagascar among African countries, continue to the present.

Reference has already been made to the proliferation of government employees which restricts the funds available for development. In 1966, for example, the total budget was $116 million, of which 87·5 per cent was for current needs and only 12·5 per cent for development expenditures.

Assistance from France is particularly important in sustaining the development budget. While the Malagasy élite admire French culture and appreciate the continuing and reliable character of aid from that country, this dependence does produce a degree of ambivalence in Franco–Malagasy relations.

Conclusion

There are, as has been seen, many factors contributing to the retarded economic development of Madagascar. They are often interconnected in complex ways and are not always easy to categorise as physical, economic, cultural or political. The two most important explanations, it is suggested, are the cultural traits associated with the 'cult of the dead' and the lack of integration in the economic geography of the island.

Education and time will doubtless change the Malagasy attitudes which are now so inhibitive to economic advance. The lack of integration presents a difficult dilemma which cannot be solved overnight. On the one hand, funds are not available to achieve the desired integration in the short run. On the other hand, the dependence on a large number of ports and the existence of an inadequate road network reduces the ability to secure these funds. The development of surface transport is hampered by the great size of the island, its rugged topography, heavy seasonal or year-round rainfall, and the wide dispersion of productive regions.

The goal should be the concentration of port activity in a limited number of ports, say Tamatave, Diégo-Suarez, Majunga and Tuléar, accompanied by the construction of roads from the other ports and commercial 'islands'. Emphasis on the Pangalanes Canal or on continued dependence on slow, dangerous and expensive coastal movements should be eliminated wherever possible. It is essential that improvement of roads and suppression of minor ports go hand in hand. This has not always been done with the elimination of installations to date; the closing of Fénérive, Vatomandry, Mahanoro, Nossi-Varika and Farafangana on the east coast, for example, has not been accompanied by a satisfactory upgrading of the east-coast road, which is in very poor condition over extensive stretches.

On the east coast, traffic from all of the ports from Maroantsetra to Farafangana could be focused on Tamatave, which would eliminate the need for coastal traffic on this most dangerous coastal stretch and add perhaps 10 per cent to the total international traffic at that port. This addition plus the growth of traffic that may be expected as the economy develops would justify enlarging the deep-water facilities at Tamatave and finally creating a large, fully protected harbour by extending the dike across the South Pass to the Grand Reef.

In the north, Antalaha, Sambava, Vohémar and Port St Louis could be tied to Diégo-Suarez. Hell Ville on Nossi-Bé must be retained to serve that productive island, but it could be eliminated as a liner stop and tied by coastal vessels either to Diégo-Suarez or to Majunga. Analalava could be made tributary to Majunga.

The remaining west-coast ports except Tuléar will probably have to continue to rely on coastal movements because of the distances separating them and the low level of their traffic. This coast is, however, more hospitable than the east coast and some of the traffic can continue to be handled by dhows, which visit this stretch in considerable numbers each year. Plans for developing some of the potentially productive river basins of the west may, furthermore, justify better road connections to the two main west-coast ports in another decade or so. Attention should be given to the creation of deep-water facilities at Majunga and to improving the rudimentary pier at Tuléar. The situation at Majunga will require careful study, since a 1-km. (1,094-yd) long dike constructed before the last war as part of a deep-water installation was rendered useless by the deposition of alluvium from the Betsiboka river before the port could be completed.

The road pattern of Madagascar would, given the above arrangements, more nearly represent a lattice than a spine with occasional ribs. Coastal roads would more or less parallel the main dorsal highway, at least southwards from Majunga on the west and along almost the full length of the east coast. The isolated commercial 'islands' of Madagascar would thus gradually be linked by land bridges. There would be a gradual expansion and coalescing of some of the 'islands'. Madagascar would eventually become an integrated island rather than an archipelago of separate and distinct productive regions.[1]

1 For further information on Madagascar, see William A. Hance, 'Transportation in Madagascar', Geographical Review, vol. 48, (1958) 45–68, and African Economic Development, 2nd (rev.) ed. (New York, 1967); Hildebert Isnard, Madagascar (Paris, 1964); Raymond Kent, From Madagascar to the Malagasy Republic (New York, 1962) and 'Madagascar Emerges from Isolation', Africa Report (Oct 1962) 3–8 f.; Paul Ottino, Les Économies Paysannes Malgaches du Bas Mangoky (Paris, 1963); République Malgache, Commissariat Général au Plan, Économie Malgache: Évolution 1950–1960 (Tananarive, June 1962) and Plan Quinquennal 1964–1968 (Tananarive, 1965); Ministère des Finances et du Commerce, Institut National de la Statistique et de la Recherche Économique, Inventaire Socio-Économique de Madagascar 1960–1965 (Tananarive, 1966); Virginia Thompson and Richard Adloff, The Malagasy Republic: Madagascar Today (Stanford, 1965). See also the journal Industries et Travaux d'Outre-Mer, and the Madagascar Bulletin Mensuel de Statistique.

Index

Major references are in bold type

Abengourou, 112, fig. 7.3
Abidjan, 7, fig. 2.1, 16, 49, 99, **103–26**, fig. 7.1, **fig. 7.2**, fig. 7.3, 147, 166, plates 5, 6; economic effects, 111–19; facilities, 11, 109–11; fishing, 115; industries, 48, 117–19, 259; origins and growth, 105–9; site, 17, 21, 22, 106, 165; traffic, 55–6, 103, 119–26
Accra, fig. 2.1, fig. 8.1, 136, 138, 139, 142, plate 7; facilities, 131, 133; industries, 259, traffic, 127
Achiasi, fig. 8.1, 136, 138
Ada, 131, fig. 8.1, 139–40
Agoué, 150, 153, 156
Agouvi, 162
Akassa, 23, 172
Akchar, 27
Akjoujt, fig. 3.1, 39
Akosombo, fig. 8.1, 139, 142
Akpakpa, 164
Algeria, 46
Andriamena, 250
Anécho, 154
Angola, 259
Antalaha, fig. 14.1, fig. 14.2, 265
Antongil Bay, 251
Antsirabe, 252, fig. 14.1, 254, fig. 14.2, 259
Apapa, 179
Arivonimamo, 256
Arusha, 240, 242
Assini, 105, 131
Aswan, fig. 12.2
Aswan High Dam, 204
Atar, fig. 3.1, 39
Atbara, 206, fig. 12.2, 217, 221
Aumale, fort, 186
Avrékété, 154
Axim, 15, 16, 18, 149
Ayamé, 117
Aythab, 204, 205
Azeffal, 27

Babanousa, 211, fig. 12.2
Badagri, 18; Creek, 179
Badi, 204

Bagamoyo, fig. 13.2, 234, 236
Baie du Repos, 32
Bangui, fig. 11.1, 199
Bas Kouilou, fig. 11.1, 185, fig. 11.2, 195
Bathurst, fig. 2.1, 15, 17, 19, 23
Bélinga, 199
Benin, 172; Bight of, 147–8, 153, 166, 173, 176; river, fig. 2.1, 17
Benue, river, 175
Betsiboka, river, 256
Betsiléo, 252
Biafra, Bight of, 176
Bissau, fig. 2.1, 19, 23
Bo, 60, fig. 5.1
Boali (Diosso), 190
Bobo-Dioulasso, 112
Bomi Hills, 75
Bong Mining Company, fig. 6.1, 97
Bonny, fig. 10.1, 172–6 *pass.*; oil terminal, 168, 173–4, 177; river, fig. 2.1, 24
Bonthe, 59, 62, fig. 5.1
Bornu, 24
Bou Lanouar, 31, 39
Bouaké, fig. 7.3, 116–18
Brass, fig. 2.1, 23, 172
Brazzaville, fig. 11.1, 187, 259
Buchanan, fig. 2.1, fig. 6.1, **fig. 6.3**, plate 4; development problems, 94, 97, 100–1; facilities, 80–1; hinterland, 77–8, 88–9, 96; traffic, 84–5, 89–90, fig. 6.6
Bukoba, 240
Burutu, 13, 23–4, 168, fig. 10.1, 173
Bushrod Island, fig. 6.2, 97

Calabar, 23–4, fig. 10.1, 172–3, 176
Cameroun, fig. 11.1, 199
Canary, Current, 30; Islands, 30, 33, 44
Cansado, fig. 3.2, 32
Cap Blanc, 28, fig. 3.2, 30
Cap Lopez, 151, 185, 187, fig. 11.2, 193–4, fig. 11.3; bay, fig. 11.2, 189, fig. 11.3
Cape Coast, fig. 2.1, 15, 131, fig. 8.1, 150
Cape Palmas, fig. 2.1, 14–16 *pass.*

Index

Cape Roxo, 15
Cape St Anne, 12, fig. 2.1, 16, 17, 19
Cape Three Points, fig. 2.1, 105, 149
Cape Verde, 12, fig. 2.1, 16, 21, 52, fig. 4.3; islands, 44
Casamance, 52
Central African Republic, fig. 11.1, 195, 197, 248
Chad, fig. 11.1, 197, 199
Choum, 39
Clay (Kle), 79
Comoros Islands, 250
Conakry, fig. 2.1, 16–19 *pass.*, 23, fig. 4.2, 51, 151
Congo–Brazzaville, **183–200**
Congo estuary, 183
Congo–Ocean Railway, fig. 11.2, 190, 195, 199
Congo, river, fig. 11.1, 189
Constitución, 165
Côte des Phoques, fig. 3.2, 30
Cotonou, 7, 11, fig. 2.1, 16, 49, **147–66**, fig. 9.1, **fig. 9.2**, plate 10; development, 150–4; economic effects, 163–4; facilities, 19–21, 154–6; site, 17, 147–50; traffic, 156–63
Cross, river, fig. 2.1, 18

Dakar, fig. 2.1, 16, 20–1, 27, fig. 3.1, 35, **41–56**, **fig. 4.1**, fig. 4.2, **fig. 4.3**, 71, 106, 151, 219, 259, plate 2; bunkering, 43–5; commercial traffic, 45–54; hinterland, 54–6
Daloa, fig. 7.3, 115
Dar es Salaam, **225–45**, fig. 13.1, fig. 13.2, fig. 13.3, 259, plate 15; facilities, 226; hinterland, 240, 242; site, 230–1; traffic, 227, 238–9, 243
Degema, fig. 10.1, 172–3
Diégo-Suarez, 251–2, fig. 14.1, fig. 14.2, 257, 264–5
Djenkin (Godomé Plage), 150
Dolisie, fig. 11.2, 195, 199
Dosso, 164
Douala, 49, fig. 11.1, 199, 259

East African seaports, **225–45**, fig. 13.1, fig. 13.2, fig. 13.3
Ebrié Lagoon, 103–9 *pass.*, fig. 7.1, fig. 7.2
Ekpé (Semé), 105, 154
El Obeid, 209, fig. 12.2, 217

Enterprise Générale d'Atlantique, 33
Enugu, 24
Escravos, river, fig. 2.1, 23

Farafangana, fig. 14.1, fig. 14.2, 256, 264
Farmington, river, 86
F'Derik (Fort Gouraud), fig. 3.1, 31, 55
Fénérive, fig. 14.1, fig. 14.2, 264
Fernand Vaz Lagoon, fig. 11.2, 189
Fianarantsoa, 252, fig. 14.1, 254, 257
Fianarantsoa–Côte Est Railway (F.C.E.), 252, fig. 14.1, 254
Firestone Rubber Company, 77, 79, 86
Flamingo Bay, 203
Flumpa, 89
Fodikwan, 203
Forcados, river, fig. 2.1
Fort Archambault, 184, 199
Fort Dauphin, fig. 14.1, fig. 14.2, 257
Fort Gouraud, *see* F'Derik
Fort Lamy, 199
Fortingaira, 66
Foulpointe, 256
Foya, 77
Franceville, fig. 11.1, 195
Freetown, fig. 2.1, 15, 19, 23, **57–73**, fig. 5.1, **fig. 5.2**; economic effects, 68–70; facilities, 23, 61–2; hinterland, 62–8, 70; site, 8, 17, 57–9; traffic, 59–61, 62

Gabon, 33, 46, **183–200**; estuary, 187, fig. 11.2, 189, 192
Gagnoa, 112, fig. 7.3, 115
Gamba, 194
Gambia, fig. 4.2, 51, 52, 54
Ganta, fig. 6.1, 77, 79, 88, 89, 96
Gash delta, 209, fig. 12.2, 217
Gba, 88
Gbanga, fig. 6.1, 77
Gbangbama, fig. 5.1, 67
German Liberian Mining Company (DELIMCO), 79; pier, fig. 6.2
Gezira irrigation scheme, 208, 209, 211, fig. 12.2
Ghana, **127–45**
Grand Bassam, 103–9 *pass.*, fig. 7.1, fig. 7.3, 149, 151
Grand Canary, 44
Grand Lahou, 105, fig. 7.3, 114
Grand Popo, 147–56 *pass.*, fig. 9.1, 163
Grebo, 91, 93

Greenville, fig. 6.1, **fig. 6.4**, 94; development problems, 93–100; facilities, 80–2; hinterland, 77–9, 90–2, 101; traffic, 84–5
Guinea, 47, 51, 77, 89; Gulf of, 7, 147, 183
Guneid, 203
Gwata, 172

Haiya, 209, fig. 12.2, 219, 223
Hangha, 62
Harbel, fig. 6.1, 77, 79
Harper, fig. 6.1; development, 100–1; facilities, 80, 82, 93; hinterland, 77–9, 91–2, 96, 101; traffic, 84–5, 88
Hell Ville, 265
Hinvi, 162
Houin, 162

Iguéla Lagoon, 189
Indian Ocean, 214, 228, 231, 232, fig. 13.2, 235, fig. 13.3
Ivato, 256

Kakata, fig. 6.1, 77
Kano, 24
Kaolack, 19, fig. 4.2, 49, 50, 52, 54
Karloke, fig. 6.1, 96
Kassala, 209, fig. 12.2
Katanga, 199
Kedia d'Idjil, 31, 39
Kenema, 67; district, 66
Keta, 131, fig. 8.1
Khartoum, fig. 12.2, 217, 221; airport, 208; North, 203
Khashm el Girba, 203, 212, fig. 12.2
Khor Arba'at, 222
Kibi, 139
Kigoma, 236
Kilindini harbour (Mombasa), 230, 235
Kilosa, 240
Kilwa, 232, fig. 13.2
Kinshasa, 259
Kisumu, 236
Kpara, 66
Kpémé, 24
Krahn-Bassa, 79, 91
Koko, fig. 2.1, 23, 168, fig. 10.1, 173, 174, 179
Kolahun, fig. 6.1, 77
Kono district, 66
Kotoku, fig. 8.1, 136, 138
Kordofan province, 209, fig. 12.2

Kouilou, river, 185, 188, 189
Kumasi, fig. 2.1, 15, 133, 136, 138, 142, 143

La Güera, fig. 3.2, 32–3
Lac Alaotra, 250, 252
Lagos, fig. 2.1, 15–16, 109, 147–9, fig. 9.1, 159, 163, 165–6, fig. 10.1, 172, 175–7; plate 12; development, 179, 181; facilities, 22; site, 21–2; traffic, 173–4
LAMCO, *see* Liberian American Swedish Minerals Company
Lam-Lam, 51
Lamu, 232, fig. 13.2
Las Palmas, 44
Lévrier bay, 27, 28, fig. 3.2, 30
Liberia, 2, 8, **75–101**
Liberia Mining Company (L.M.C.), 75, 79, 93; pier, fig. 6.2
Liberian Agricultural Company, 88, 96
Liberian American Swedish Minerals Company (LAMCO), 77, 81, 88, 89, 90, 94
Libreville, 151, 183, fig. 11.1, **fig. 11.2**, 190, **192–5**, 198–9, 200–1; facilities, 192–3; origins and growth, 186–7; traffic, 187, 189, 192, 194, 200
Lindi, fig. 13.2, 236
Loango, 151, 185, fig. 11.2, 190
Lomé, fig. 2.1, 16, 19, 20, 21, 49, fig. 9.1, 149, 154, 165–6
Lunsar, 24

Madagascar, 8, **247–65**
Madeira, 44
Mahanoro, fig. 14.1, fig. 14.2, 264
Mahavavy, 257
Mahendy, 165
Majunga, 252, fig. 14.1, fig. 14.2, 257, 264, 265
Makeni, fig. 5.1
Makokou-Mékambo, fig. 11.1, 193
Mali, 41, fig. 4.2, 48, 51, 52, 54, 55, 56
Malindi, 232, fig. 13.2
Man, fig. 7.3, 115, 116
Managil extension, 211, fig. 12.2, 219
Manakara, 252, fig. 14.1, 257
Mandji, *see* Port Gentil
Mandraka, 250
Mandrare valley, 257
Mano, river, 97
Maradi, 159

Marampa, 63, fig. 5.1, 64, 66
Marsa Kuwai, 221
Marsa Sheikh Bargout (Port Sudan), 205, 221
Marshall, 86
Maryland county, 96
Massabi, 185, fig. 11.2
Matadi, 151, fig. 11.1, 190
Mauritania, 27, 41, fig. 4.2, 51–6 *pass.*
Mauritius, 260
Mayoumba, fig. 11.1, 185, 189, 194
M'Bao, 46, fig. 4.3, 55
Mensurado bridge, 95
MIFERMA, *see* Société des Mines de Fer de Mauritanie
Mikindani, fig. 13.2, 236
Moanda, 195, 200
Mogadishu, 234
Mokanji Hills, fig. 5.1, 67
Mombasa, 219, **225–45**, fig. 13.1, fig. 13.2, fig. 13.3, plate 14; facilities, 226; hinterland, 236, 240, 242; site, 230–1; traffic, 227, 243
Mondah bay, 189
Monrovia, fig. 2.1, 16, 21, 75, fig. 6.1, **fig. 6.2**, 97, 99, plate 3; development, 100–1; facilities, 79–80, 94–6; hinterland, 77–8; traffic, 84–9
Montserrado county, 77
Moroantsetra, fig. 14.1, fig. 14.2, 264
Moshi, 236, 240, 242
Mount Coffee, 85
Mozambique, 232, 259; channel, 250
Mtwara, **225–45**, fig. 13.2, fig. 13.3; facilities, 226, 238; hinterland, 240; site, 230–1; traffic, 227, 243
Mueller, William M., & Co., 75
Musoma, 240
Mwanza, 240

Nairobi, 259
National Iron Ore Company (NIOC), 79, fig. 6.2
New Cess, 96
Niger delta, 168, 178; river, 175
Nigeria, **167–82**
Nigerian Ports Authority, 181
Nimba, fig. 6.1, 77, 81, 88; county, 88
Nokoué, Lake, 163, 164
Nossi-Bé, fig. 14.1, fig. 14.2, 257, 265
Nossi-Varika, fig. 14.1, 264
Nouadhibou (Port Étienne), 7, **27–40**, fig. 3.1, **fig. 3.2**, fig. 4.2, 54–6 *pass.*,

plate 1; fishing, 30, 33–5; growth, 30–3; site, 28–30; traffic, 32, 35–40
Nouakchott, 27, fig. 3.1, 28, 31, 39, fig. 4.2, 56
Nyala, 211, fig. 12.2, 217
Nyika, 230
N'Zérékoré, 77

Ode Itsekiri, 172
Office de Commercialisation Agricole (Senegal), 52
Ogooué, river, fig. 11.1, 185, 187, fig. 11.2, 189, 200
Okrika, fig. 10.1, 169, 172, 174, 177, 178
Opération Hirondelle, 159
Opobo (Egwenga), 173
Organisation Commune Dahomey–Niger (O.C.D.N.), 164
Oseif, 203, fig. 12.2
Ouagadougou, 112, 159
Ouémé, river, 185
Ouidah (Whydah), 12, 18, 148, fig. 9.1, 150, 151, 154, 156, 163, 173
Owendo, fig. 11.1, fig. 11.2, 192, 193, 199, 200

Pangalanes canal, fig. 14.1, 256, 264
Pangani, fig. 13.2, 234, 236
Parakou, 159, 164
Pate, 232, fig. 13.2
Pendembu, 59, fig. 5.1
Peninsula Mountains, 57, 58
Pepel, fig. 2.1, 24, fig. 5.1, 64, **fig. 5.2**, 65, 66, 68
Point Central, 37
Point Sam, fig. 5.1, 67
Pointe Indienne, fig. 11.2, 189, 195
Pointe Noire, 183, fig. 11.1, **fig. 11.2**, 189, **fig. 11.3, 195–8**, plate 13; facilities, 190–1, 198; fishing, 197; growth, 190–1; hinterland, 190, 197, 199; traffic, 185, 191, 195–6, 198–200
Port Bouet, fig. 2.1, 19, 103, fig. 7.1, 106, 108, 114, 149
Port Étienne, *see* Nouadhibou
Port Gentil, 183, fig. 11.1, **fig. 11.2**, 190, **192–4, fig. 11.3**, 198–9; facilities, 193, 200; growth, 187; industries, 201; traffic, 187, 189, 191, 193–4
Port Harcourt, fig. 2.1, 13, 16, 24, fig. 10.1, 169–81 *pass.*, plate 11
Port St Louis, fig. 14.1, fig. 14.2, 265

Port Sudan, **203–23, fig. 12.1,** fig. 12.2; development problems, 214–23; facilities, 208–9, 217–18; hinterland, 217, 219–21; origins and growth, 204–8; site, 205–7; traffic, 208–16
Porto Novo, 156, 173

Rahad, 211, fig. 12.2
Red Sea, 204, 207, fig. 12.2, 217, 231
Réunion, 250
Rio de Oro (Spanish Sahara), 28, fig. 3.1, fig. 3.2, 32
Rio Muni bay, 189
River Cess, fig. 6.1, 85, 92
Robertsport, fig. 6.1; development, 92, 100; hinterland, 77–9; traffic, 85
Roseires, 211, fig. 12.2, 219
Rovuma, river, 234
Rufisque, 52, fig. 4.3
Russwurm Island, 92

Safi, 165
St Louis du Senegal, 15, 16, fig. 3.1, fig. 4.2, 51, 52
Saloum, 206, 207
Saltpond, 131, fig. 8.1
Sambirano, 257
San Pedro, 21, fig. 7.1, 105, 126
Santo Domingo, 151
Sapele, fig. 2.1, 23, 168, fig. 10.1, 172, 173
Sassandra, 49, 103, fig. 7.1, 105, 112, fig. 7.3
Sefadu, fig. 5.1, 66
Sekondi, fig. 2.1, 16, 19, 20
Senegal, river, 49
Senegambia, 51, 54
Sennar, 209, fig. 12.2
Setté Cama, 185, 189
Sherbro island, 67; river, 59
Sierra Leone Government Railway, 60, 64, 68, 72
Sierra Leone Selection Trust, 66
Société des Mines de Fer de Mauritanie (MIFERMA), 31–8 *pass.*
Société Industrielle de la Grande Pêche (SIGP), 33
Société Mauritanienne d'Accorage et de Manutention (SAMMA), 35–7 *pass.*
Société Mauritanienne de Pêche (SOMAP), 34
Société Miniere de Mauritanie (SOMIMA), 39
Sofala, 232

Somalia, 262
Soubré, fig. 7.3, 115
Spanish Sahara, *see* Rio de Oro
Suakin, 204–7 *pass.*, fig. 12.2, 221
Suez Canal, 44, 46, 56, 214, 217–18, 235, 262

Tabou, 103, fig. 7.1, 105, fig. 7.3, 149
Taiba, 46, 51–2
Takoradi, 11, fig. 2.1, 19, 21, **127–38,** fig. 8.1, **fig. 8.2,** 140, 142, 165, plate 8; facilities, 20, 133–5; industries, 143; traffic, 127–37 *pass.*, 143–5
Tananarive, 8, 250–9 *pass.*, fig. 14.1, fig. 14.2
Tananarive–Côte Est Railway (T.C.E.), 252, fig. 14.1, 254
Tanga, **225–45,** fig. 13.1, fig. 13.2, fig. 13.3; facilities, 226, 243; hinterland, 240, 242; site, 230–1; traffic, 227
Tamatave, 250–9 *pass.*, fig. 14.1, fig. 14.2, 264
Tappita, fig. 6.1, 88
Tarkwa, 16, fig. 8.1
Tatuke, 96
Tema, fig. 21, fig. 8.1, **139–45, fig. 8.3,** 147, 165–6, plate 9; facilities, 20; industries, 142–3; site, 139–40; traffic, 130, 144–5
Tenerife, 44
Thiès, 46, fig. 4.2, 51
Tivaouane, fig. 4.2, 52
Togo, 8
Tokar delta, fig. 12.2, 217
Tongo, fig. 5.1, 66
Towartit, 217
Trou sans fond, 106, fig. 7.2, 108
Tuléar, 251–7 *pass.*, fig. 14.1, fig. 14.2, 264–5

Upper Volta, 47, 51

VALCO, *see* Volta Aluminium Company
Vatomandry, fig. 14.1, fig. 14.2, 264
Victoria, Lake, 237, fig. 13.3
Vohémar, fig. 14.1, fig. 14.2, 265
Volta Aluminium Company (VALCO), fig. 8.3, 142, plate 9
Volta, river, fig. 2.1, 17, 138–40
Vridi, canal, 22, 103–10 *pass.*, fig. 7.1, fig. 7.2, 115

Wadi Halfa, 204, 207, fig. 12.2
Warri, fig. 2.1, 23, 168, fig. 10.1, 172, 173
Wau, 211, fig. 12.2
Winneba, 131, fig. 8.1

Yaoundé, fig. 11.1, 199
Yekepa, fig. 6.1
Yengema, fig. 5.1, 66
Yila, fig. 6.1, 88
York Island, 59

Zambia, 238–9, fig. 13.3, 242, 245
Zanaga, 199
Zanzibar, 225–45, fig. 13.1, fig. 13.2;
  facilities, 226; hinterland, 238; site,
  230–2
Ziguinchor, fig. 2.1, 23, fig. 4.2, 50, 52,
  54
Zinder, 159
Zouérate, fig. 3.1, 31, 39
Zwedru, fig. 6.1, 79, 91, 93